# Lecture Notes in Control and Information Sciences 354

Editors: M. Thoma, M. Morari

Marcin Witczak

# Modelling and Estimation Strategies for Fault Diagnosis of Non-Linear Systems

## From Analytical to Soft Computing Approaches

Springer

**Author**
Dr. Marcin Witczak
Institute of Control and Computation Engineering
University of Zielona Góra
Ul. Podgórna 50
65-246 Zielona Góra
Poland
E-mail: M.Witczak@issi.uz.zgora.pl



Library of Congress Control Number: 2007922361

ISSN print edition: 0170-8643
ISSN electronic edition: 1610-7411
ISBN-10 3-540-71114-7 Springer Berlin Heidelberg New York
ISBN-13 978-3-540-71114-8 Springer Berlin Heidelberg New York



Springer is a part of Springer Science+Business Media
springer.com


Typesetting: by the authors and SPS using a Springer LaTeX macro package

Printed on acid-free paper SPIN: 11959908 89/SPS 5 4 3 2 1 0

To my beloved family:
parents Danuta and Ryszard,
wife Ewa and son Maksymilian

# Preface

It is well known that there is an increasing demand for modern systems to become safer and reliable. This real world's development pressure has transformed fault diagnosis, initially perceived as the art of designing a satisfactorily safe system, into the modern science that it is today.

Indeed, the classic way of fault diagnosis boils down to controlling the limits of single variables and then using the resulting knowledge for fault-alarm purposes. Apart from the simplicity of such an approach, the observed increasing complexity of modern systems necessitates the development of new fault diagnosis techniques. Such a development can only be realised by taking into account the information hidden in all measurements. One way to tackle such a challenging problem is to use the so-called model-based approach. Indeed, the application of an adequate model of the system being supervised is very profitable with respect to gaining the knowledge regarding its behaviour. A further and deeper understanding of the current system behaviour can be achieved by implementing parameter and state estimation strategies. The obtained estimates can then be used for supporting diagnostic decisions.

Although the majority of industrial systems are non-linear in their nature, the most common approach to settle fault diagnosis problems is to use well-known tools for linear systems, which are widely described and well documented in many excellent monographs and books. On the other hand, publications on fault diagnosis for non-linear systems are scattered over many papers and a number of book chapters.

Taking into account the above-mentioned conditions, the book presents selected modelling and estimation strategies for fault diagnosis of non-linear systems in a unified framework. In particular, starting from the classic parameter estimation techniques through advanced state estimation strategies up to modern soft computing, the discrete-time description of the system is employed. Such a choice is dictated by the fact that the discrete-time description is easier and more natural to implement on modern computers then its continuous-time counterpart. This is especially important for practicing engineers, who usually are not fluent in complex mathematical descriptions.

The book results from my research in the area of fault diagnosis for non-linear systems that has been conducted since 1998. The only exception is Part I, which presents general principles of fault diagnosis and outlines frequently used approaches. The remaining part of the book is organised as follows: Part II presents original research results regarding state and parameter estimation-based fault diagnosis. Part III is devoted to the so-called soft computing techniques. In particular, a number of original fault diagnosis schemes that utilise either evolutionary algorithms or neural networks are presented and carefully described.

This book is primarily a research monograph which presents, in a unified framework, some recent results on modelling and estimation techniques for fault diagnosis of non-linear systems. The book is intended for researchers, engineers and advanced postgraduate students in control and electrical engineering, computer science as well as mechanical and chemical engineering.

Some of the research results presented in this book were developed with the kind support of the Ministry of Science and Higher Education in Poland under the grant 4T11A01425 *Modelling and identification of non-linear dynamic systems in robust diagnostics.* The work was also supported by the EC under the RTN project (RTN-1999-00392) *DAMADICS.*

I would like to express my sincere gratitude to my family for their support and patience. I am also grateful to Prof. Józef Korbicz for suggesting the problem, and for his continuous help and support. I also would like to express my special thanks to Mr Przemysław Prętki and Dr Wojciech Paszke for their help in preparing some of the computer programmes and simulations of Part III and II, respectively. I am very grateful to Ms Agnieszka Rożewska for proofreading and linguistic advise on the text. Finally, I wish to thank all my friends and colleagues at the Institute of Control and Computation Engineering of the University of Zielona Góra, who helped me in many, many ways while I was preparing the material contained in this book.

October 2006 *Marcin Witczak*

Zielona Góra

# List of Symbols

| | |
|---|---|
| $t$ | time |
| $k$ | discrete time |
| $\mathcal{E}(\cdot)$ | expectation operator |
| $\mathrm{Co}(\cdot, \ldots, \cdot)$ | convex hull |
| $\underline{\sigma}(\cdot)$, $\bar{\sigma}(\cdot)$ | minimum, maximum singular value |
| $\mathcal{N}(\cdot, \cdot)$ | normal distribution |
| $\mathcal{U}(\cdot, \cdot)$ | uniform distribution |
| $\boldsymbol{A} \succ (\succeq)\mathbf{0}$ | positive (semi-)definite matrix |
| $\boldsymbol{A} \prec (\preceq)\mathbf{0}$ | negative (semi-)definite matrix |
| $\rho(\boldsymbol{A})$ | spectra radius of $\boldsymbol{A}$ |
| $\boldsymbol{x}_k, \hat{\boldsymbol{x}}_k$ $(\dot{\boldsymbol{x}}(t), \dot{\hat{\boldsymbol{x}}}(t)) \in \mathbb{R}^n$ | state vector and its estimate |
| $\boldsymbol{y}_k, \hat{\boldsymbol{y}}_k \in \mathbb{R}^m$ | output vector and its estimate |
| $\boldsymbol{y}_{M,k} \in \mathbb{R}^m$ | model output |
| $\boldsymbol{e}_k \in \mathbb{R}^n$ | state estimation error |
| $\boldsymbol{\varepsilon}_k \in \mathbb{R}^m$ | output error |
| $\boldsymbol{u}_k \in \mathbb{R}^r$ | input vector |
| $\boldsymbol{d}_k \in \mathbb{R}^q$ | unknown input vector, $q \leq m$ |
| $\boldsymbol{z}_k \in \mathbb{R}^m$ | residual |
| $\boldsymbol{r}_k \in \mathbb{R}^{n_p}$ | regressor vector |
| $\boldsymbol{w}_k$, $\boldsymbol{v}_k$ | process and measurement noise |
| $\boldsymbol{Q}_k$, $\boldsymbol{R}_k$ | covariance matrices of $\boldsymbol{w}_k$ and $\boldsymbol{v}_k$ |

| | |
|---|---|
| $\boldsymbol{p} \in \mathbb{R}^{n_p}$ | parameter vector |
| $\boldsymbol{f}_k \in \mathbb{R}^s$ | fault vector |
| $\boldsymbol{g}(\cdot)$, $\boldsymbol{h}(\cdot)$ | non-linear functions |
| $\boldsymbol{E}_k \in \mathbb{R}^{n \times q}$ | unknown input distribution matrix |
| $\boldsymbol{L}_{1,k}$, $\boldsymbol{L}_{2,k}$ | fault distribution matrices |
| $n_{1,y}, \ldots, n_{m,y}, n_{1,u}, \ldots, n_{m,u}$ | maximum lags in outputs and inputs |
| $n_t$, $n_v$ | number of input-output measurements for identification and validation |
| $n_{\text{pop}}$ | number of populations |
| $n_m$ | population size |
| $n_d$ | initial depth of trees |
| $n_s$ | tournament population size |
| $n_h$ | number of hidden neurons |
| $\mathbb{T}$, $\mathbb{F}$ | terminal and function sets |

## Abbreviations

| | |
|---|---|
| ADC | Anolog to Digital Converter |
| AFD | Active Fault Diagnosis |
| AIC | Akaike Information Criterion |
| ANN | Artificial Neural Network |
| ARS | Adaptive Random Search |
| BEA | Bounded-Error Approach |
| D-OED | D-Optimum Experimental Design |
| EA | Evolutionary Algorithm |
| EKF | Extended KF |
| ESSS | Evolutionary Search with Soft Selection |
| EUIO | Extended UIO |
| FDI | Fault Detection and Isolation |
| FIM | Fisher Information Matrix |
| FPC | Final Prediction Error |
| FTC | Fault-Tolerant Control |
| GA | Genetic Algorithm |
| GP | Genetic Programming |
| GMDH | Group Method of Data Handling |
| KF | Kalman Filter |
| LSM | Least-Square Method |
| MISO | Multi-Input Single-Output |
| MIMO | Multi-Input Multi-Output |
| MCE | Monte Carlo Evaluation |
| MLP | Multi-Layer Perceptron |
| OED | Optimum Experimental Design |
| PFD | Passive Fault Diagnosis |
| RBF | Radial Basis Function |
| RLS | Recursive Least-Square |
| SISO | Single-Output Single-Input |
| UIF | Unknown Input Filter |
| UIO | Unknown Input Observer |

# Contents

## Part II. State and Parameter Estimation Strategies



## Part III. Soft Computing Strategies

# 1. Introduction

A continuous increase in the complexity, efficiency, and reliability of modern industrial systems necessitates a continuous development in the control and fault diagnosis [16, 27, 84, 98, 96] theory and practice. These requirements extend beyond the normally accepted safety-critical systems of nuclear reactors, chemical plants or aircrafts, to new systems such as autonomous vehicles or fast rail systems. An early detection and maintenance of faults can help avoid system shutdown, breakdowns and even catastrophes involving human fatalities and material damage. A modern control system that is able to tackle such a challenging problem is presented in Fig. 1.1 [181]. As can be observed, the controlled system

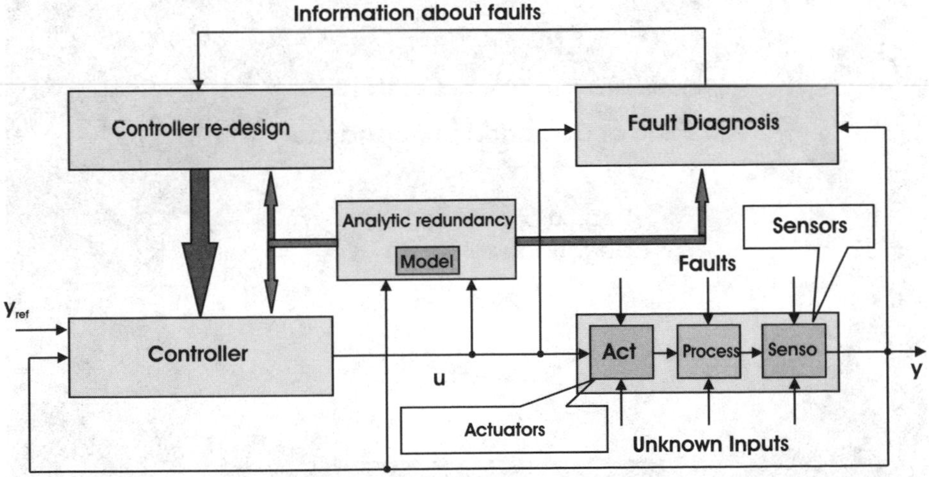


**Fig. 1.1.** Modern control system

is the main part of the scheme, and it is composed of actuators, process dynamics and sensors. Each of these parts is affected by the so-called unknown inputs, which can be perceived as process and measurement noise as well as external

M. Witczak: Model. and Estim. Strat. for Fault Diagn. of Non-Linear Syst. LNCIS 354, pp. 1–7, 2007.
springerlink.com 

disturbances acting on the system. When model-based control and analytical redundancy-based fault diagnosis are utilised [16, 27, 96], then the unknown input can also be extended by model uncertainty, i.e., the mismatch between the model and the system being considered.

The system may also be affected by faults. A fault can generally be defined as an unpermitted deviation of at least one characteristic property or parameter of the system from the normal condition, e.g., a sensor malfunction. All the unexpected variations that tend to degrade the overall performance of a system can also be interpreted as faults. Contrary to the term *failure*, which suggests a complete breakdown of the system, the term *fault* is used to denote a malfunction rather than a catastrophe. Indeed, *failure* can be defined as a permanent interruption of the system ability to perform a required function under specified operating conditions. This distinction is clearly illustrated in Fig. 1.2. Since a system can be split into three parts (Fig. 1.1), i.e., actuators, the process, and sensors, such a decomposition leads directly to three classes of faults:

- *Actuator faults*, which can be viewed as any malfunction of the equipment that actuates the system, e.g., a malfunction of the electro-mechanical actuator for a diesel engine [15]. This kind of faults can be divided into three categories:

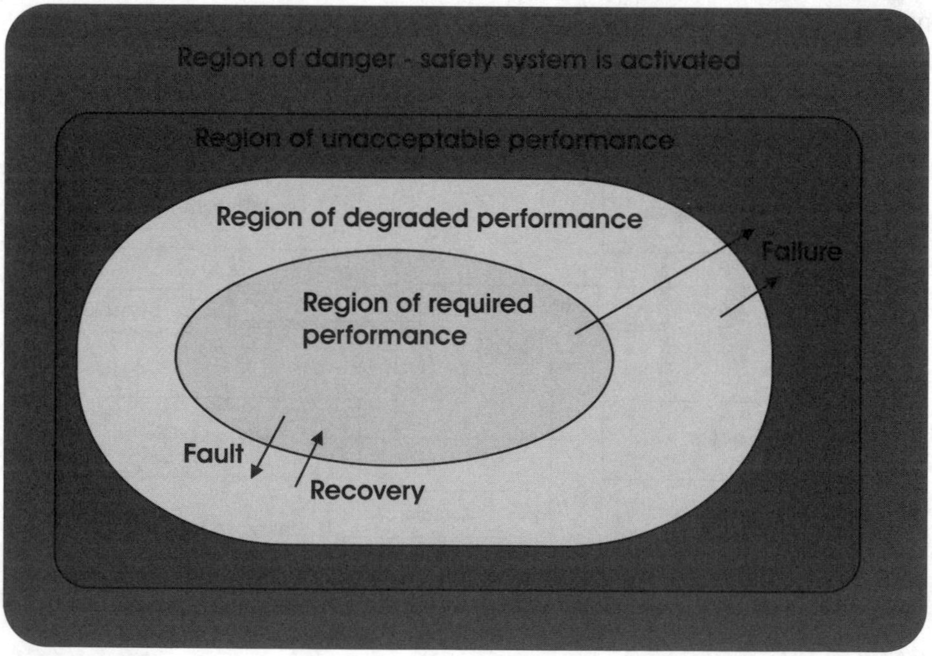


**Fig. 1.2.** Regions of system performance

Lock-in-place: the actuator is locked in a certain position at an unknown time $t_f$ and does not respond to subsequent commands:

$$u_{i,k} = u_{i,t_f} = \text{const}, \quad \forall k > t_f. \tag{1.1}$$

Outage: the actuator produces zero force and moment, i.e., it becomes ineffective:

$$u_{i,k} = 0, \quad \forall k > t_f. \tag{1.2}$$

Loss of effectiveness: a decrease in the actuator gain that results in a deflection that is smaller than the commanded position:

$$u_{i,k} = k_i u^c_{i,k}, \quad 0 < k_i < 1 \quad \forall k > t_f, \tag{1.3}$$

where $u^c_{i,k}$ stands for the required actuation;

- *Process faults* (or component faults), which occur when some changes in the system make the dynamic relation invalid, e.g., a leak in a tank in a two-tank system;
- *Sensor faults*, which can be viewed as serious measurements variations. Similarly to actuator faults, two sensor fault scenarios can be considered:

Lock-in-place: the sensor is locked in a certain position at an unknown time $t_f$ and does not provide the current value of the measured variable:

$$y_{i,k} = y_{i,t_f} = \text{const}, \quad \forall k > t_f. \tag{1.4}$$

Loss of measurement accuracy: a degradation of the measurement accuracy of the sensor:

$$y_{i,k} = k_i y^c_{i,k}, \quad \forall k > t_f, \tag{1.5}$$

while $y^c_{i,k}$ stands for the true value of the measured variable and $k_i$ is significantly different from 0.

The role of the fault diagnosis part is to monitor the behaviour of the system and to provide all possible information regarding the abnormal functioning of its components. As a result, the overall task of fault diagnosis consists of three subtasks [27](Fig. 1.3):

Fault detection: to make a decision regarding the system stage – either that something is wrong or that everything works under the normal conditions;
Fault isolation: to determine the location of the fault, e.g., which sensor or actuator is faulty;
Fault identification: to determine the size and type or nature of the fault.

However, from the practical viewpoint, to pursue a complete fault diagnosis the following three steps have to be realised [56]:

Residual generation: generation of the signals that reflect the fault. Typically, the residual is defined as a difference between the outputs of the system and its estimate obtained with the mathematical model;

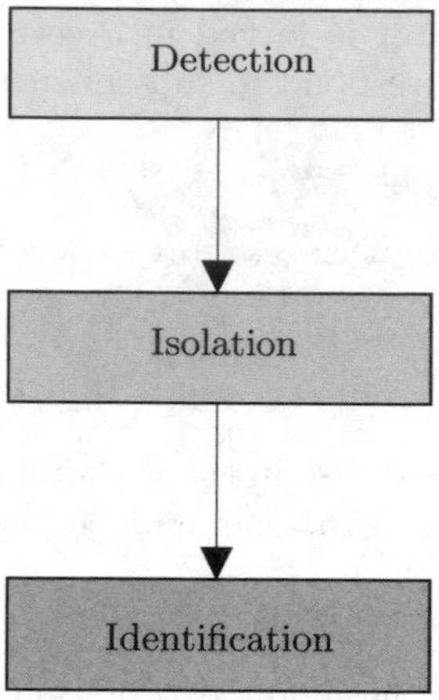


**Fig. 1.3.** Three-stage process of fault diagnosis

Residual evaluation: logical decision making on the time of occurrence and the location of faults;
Fault identification: determination of the type of a fault, its size and cause.

The knowledge resulting from these steps is then provided to the controller redesign part, which is responsible for changing the control law in such a way as to maintain the required system performance. Thus, the scheme presented in Fig. 1.1 can be perceived as a fault-tolerant one.

Fault-Tolerant Control (FTC) [16] is one of the most important research directions underlying contemporary automatic control. FTC can also be perceived as an optimised integration of advanced fault diagnosis [96, 179] and control [16] techniques.

Finally, it is worth to note that the word *symptom* denotes a change of an observable quantity from normal behaviour.

The main objective of this part is to present a general description of the state-of-the-art regarding analytical and soft computing-based Fault Detection and Isolation (FDI). It should also be noted that the material of this part is limited to the approaches that can be extended or directly applied to non-linear systems.

## 1.1 Introductory Background

If residuals are properly generated, then fault detection becomes a relatively easy task. Since without fault detection it is impossible to perform fault isolation and, consequently, fault identification, all efforts regarding the improvement of residual generation seem to be justified. This is the main reason why the research effort of this book is oriented towards fault detection and especially towards residual generation.

There have been many developments in model-based fault detection since the beginning of the 1970s, regarding both the theoretical context and the applicability to real systems (see [27, 96, 138] for a survey). Generally, the most popular approaches can be split into three categories, i.e.,

- parameter estimation;
- parity relation;
- observer-based.

All of them, in one way or another, employ a mathematical system description to generate the residual signal. Except for parameter estimation-based FDI, the residual signal is obtained as a difference between the system output and its estimate obtained with its model, i.e.,

$$\boldsymbol{z}_k = \boldsymbol{y}_k - \hat{\boldsymbol{y}}_k. \tag{1.6}$$

The simplest model-based residual generation scheme can be realised in a way similar to that shown in Fig. 1.4. In this case, the design procedure reduces to system identification, and fault detection boils down to checking the norm of the residual signal $\|\boldsymbol{z}_k\|$. In such a simple residual generation scheme, neural networks seem to be especially popular [96, 139].

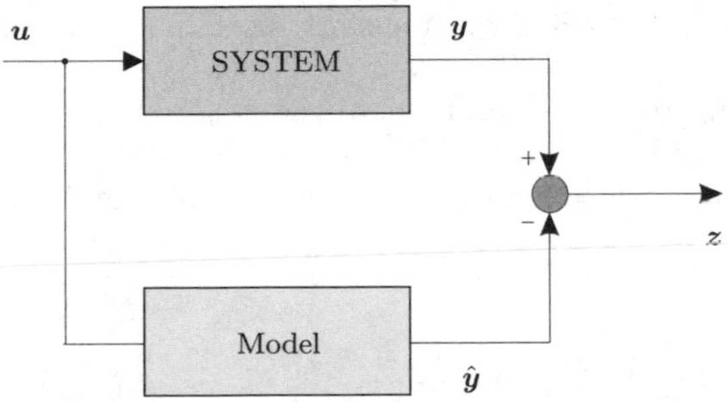


**Fig. 1.4.** Simple residual generation scheme

Irrespective of the identification metod used, there is always the problem of model uncertainty, i.e., the model-reality mismatch. Thus, the better the model used to represent system behaviour, the better the chance of improving the reliability and performance in diagnosing faults. This is the main reason why the fault detection scheme shown in Fig. 1.4 is rarely used for maintaining fault diagnosis of high-safety systems. Indeed, disturbances as well as model uncertainty are inevitable in industrial systems, and hence there exists a pressure creating the need for robustness in fault diagnosis systems. This robustness requirement is usually achieved at the fault detection stage, i.e., the problem is to develop residual generators which should be insensitive (as far as possible) to model uncertainty and real disturbances acting on a system while remaining sensitive to faults. In one way or another, all the above-mentioned approaches can realise this requirement for linear systems.

Other problems arise from fault detection of non-linear systems. Indeed, the available non-linear system identification techniques limit the application of fault detection. For example, in the case of observer-based FDI non-linear state-space models cannot be usually obtained using physical considerations (physical laws governing the system being studied). Such a situation is usually caused by the high complexity of the system being considered. This means that a model which merely approximates system-input behaviour (no physical interpretation of the state vector or parameters) should be employed.

The process of fault isolation requires usually more complex schemes than the one of fault detection. Indeed, this very important task of FDI is typically realised by either the so-called dedicated or generalised schemes [27, 84]. In the case of the dedicated scheme, residual generators are designed in such a way that each residual $z_i$, $i = 1, \ldots, s$ is sensitive to one fault only while it remains insensitive to others. Apart from a very simple fault isolation logic, which is given by

$$|z_{i,k}| > T_i \quad \Rightarrow f_{i,k} \neq 0, \quad i = 1, \ldots, s, \tag{1.7}$$

where $T_i$ is a predefined threshold, this fault isolation design procedure is usually very restrictive and does not allow achieving additional design objectives such as robustness to model uncertainty. Irrespective of the above difficulties, the dedicated fault isolation strategy is frequently used in neural network-based FDI schemes [96].

On the contrary, residual generators of the generalised scheme are designed in such a way as each residual $z_i$, $i = 1, \ldots, s$ is sensitive to all but one faults. In this case, the fault detection logic is slightly more complicated, i.e.,

$$\left.\begin{array}{l} |z_{i,k}| > T_i \\ |z_{j,k}| > T_j, j = 1, \ldots, i-1, i+1, \ldots, s \end{array}\right\} \Rightarrow f_{i,k} \neq 0, \quad i = 1, \ldots, s, \tag{1.8}$$

but it requires a less restrictive design procedure than the dedicated scheme, and hence the remaining design objectives can usually be accomplished.

As has already been mentioned, the material presented in this book is limited to discrete-time systems, and hence all the approaches that cannot be applied to this class of systems are omitted.

## 1.2 Content

The remaining part of this book is divided into three main parts:

Principles of fault diagnosis: This part is composed of two chapters. Chapter 2 presents the most popular analytical approaches [27, 84, 96] that can be used for fault diagnosis. In particular, three main approaches are discussed, i.e., parameter estimation, the parity relation and observers. The application of the above approaches is discussed for both linear and non-linear systems.

Finally, the main advantages and drawbacks of the discussed techniques are portrayed. Chapter 3 presents soft computing techniques [96, 140, 151, 153, 154, 165] that can be used for fault diagnosis[96, 95, 97, 140, 181]. The attention is limited to the so-called quantitative soft computing approaches. In particular, neural networks and evolutionary algorithms [96, 140, 151, 181] are considered. The chapter portrays a short introduction to these computational intelligence methods and presents a bibliographical review regarding their application to fault diagnosis. Similarly as in Chapter 2, the advantages and drawbacks of the soft computing techniques being considered are discussed.

State and parameter estimation strategies: This part is also composed of two chapters. Chapter 4 presents original observer-based solutions [183, 184, 186, 187] that can be used for fault diagnosis of non-linear systems. In particular, this chapter is concerned with three different observer structures that are based on the general idea of an unknown input observer [27]. Chapter 5 introduces the idea of an active fault diagnosis approach. The general principle of such an approach is to actively use the input signal to enhance the knowledge about the current system behaviour. In the context of this technique, the theory of experimental design for parameter estimation of non-linear systems is introduced [7, 167, 170]. The remaining part of this chapter shows how to employ such a strategy for the estimation and fault detection of an impedance [180].

Soft computing strategies: Similarly as in the case of the preceding part, this one is composed of two chapters, too. Chapter 6 presents original evolutionary algorithm-based approaches to fault diagnosis [114, 181, 183, 187]. In particular, it is shown how to use evolutionary algorithms for model design. Another development presented in this chapter concerns the application of the evolutionary algorithms to increasing the convergence rate of the selected observers presented in Chapter 4. Chapter 7 presents also original developments regarding neural network-based robust fault diagnosis [181, 182, 185, 188]. One objective of this chapter is to show how to describe modelling uncertainty of neural networks. Another objective is to show how to use the resulting knowledge about model uncertainty for robust fault detection. It is also worth noting that this chapter presents experimental design strategies that can be used for decreasing model uncertainty of neural networks [182, 188].

# Part I

# Principles of Fault Diagnosis

# 2. Analytical Techniques-Based FDI

The main objective of this chapter is to present the main principles regarding fault detection and isolation with analytical techniques. The chapter is organised as follows: Section 2.1 presents the most popular approaches that are used for FDI of linear systems. Starting from parameter estimation strategies through the parity relation up to observers, general principles are described and commonly used algorithms are outlined. A similar scenario is implemented in Section 2.2, with non-linear systems considered instead. Section 2.3 is devoted to the robustness issues in modern FDI.

## 2.1 Approaches to Linear Systems

The main objective of this section is to present an introduction to the most popular FDI approaches for linear systems. In particular, each of them is presented and then its application to residual generation is discussed. Finally, it is shown how to use them for fault isolation.

### 2.1.1 Parameter Estimation

As has already been mentioned parameter estimation is one of the three most popular ways of FDI (see [84]). This section presents general rules of parameter estimation-based FDI for linear-in-parameter systems.

Let us consider a linear-in-parameter system described by

$$y_k = \boldsymbol{r}_k^T \boldsymbol{p} + v_k, \tag{2.1}$$

where $\boldsymbol{r}_k$ is the so-called regressor vector, e.g., $\boldsymbol{r}_k = [y_{k-1}, y_{k-2}, u_k, u_{k-1}]^T$.

Assuming that the parameter vector $\boldsymbol{p}$ has physical meaning, the task consists in detecting faults in a system by measuring the input $\boldsymbol{u}_k$ and the output $\boldsymbol{y}_k$, and then estimating the parameters of the model of the system (Fig. 2.1). Thus, the fault can be modelled as an additive term acting on the parameter vector of the system, i.e.,

$$\boldsymbol{p} = \boldsymbol{p}_{\text{nom}} + \boldsymbol{f}, \tag{2.2}$$

M. Witczak: Model. and Estim. Strat. for Fault Diagn. of Non-Linear Syst. LNCIS 354, pp. 11–30, 2007.
springerlink.com 

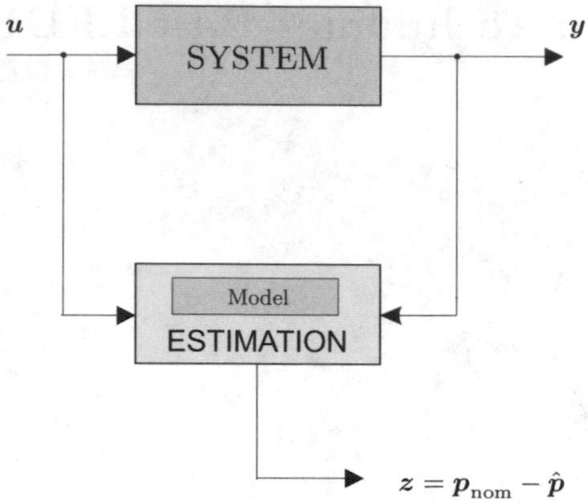


**Fig. 2.1.** Principle of parameter estimation-based residual generation

with $\boldsymbol{p}_{\text{nom}}$ standing for the nominal (fault-free) parameter vector. Therefore, the problem boils down to on-line parameter estimation, which can be solved with various recursive algorithms, such as the celebrated recursive least-square method [170], the instrumental variable approach [161] or the bounded-error approach [116]. Irrespective of the parameter estimation method employed, the fault detection logic boils down to checking if the residual norm is greater than a predefined threshold, i.e.,

$$\|\boldsymbol{z}_k\| > T \quad \Rightarrow \quad \boldsymbol{f}_k \neq \boldsymbol{0}, \tag{2.3}$$

and $\boldsymbol{z}_k = \boldsymbol{p}_{\text{nom}} - \hat{\boldsymbol{p}}_k$, where $\hat{\boldsymbol{p}}_k$ is the parameter estimate.

The main drawback of this approach is that the model parameters should have physical meaning, i.e., they should correspond to the parameters of the system. In such situations, the detection and isolation of faults is very straightforward, i.e., it is given by

$$|z_{i,k}| > T_i \quad \Rightarrow \quad f_{i,k} \neq 0, \quad i = 1, \ldots, s = n_p. \tag{2.4}$$

If this is not the case, it is usually difficult to distinguish a fault from a change in the parameter vector $\boldsymbol{p}$ resulting from time-varying properties of the system. Moreover, the process of fault isolation may become extremely difficult because model parameters do not uniquely correspond to those of the system. It should also be pointed out that the detection of faults in sensors and actuators is possible but rather complicated [138]. Indeed, sensor and/or actuator faults may influence the input and output data in the same way as the process (parameter) faults. Indeed, in [169] the authors proposed an FDI approach for sensors and actuators based on the parameter estimation framework. In particular, they employed the Extended Kalman Filter (EKF) for estimating parameter values associated with sensor and actuator faults.

Finally, it should be pointed out that the typical limitation regarding parameter estimation-based approaches is related to the fact that the input signal should be persistently exciting [67]. This condition is satisfied if the input signal $\boldsymbol{u}_k$ provides enough information to estimate $\boldsymbol{p}$. Many industrial systems, however, may not allow feeding such (persistently exciting) signals as inputs.

### 2.1.2 Parity Relation

In the case of linear systems, the following state-space description (it is possible to use different descriptions [65, 84]) can be employed (in a deterministic configuration):

$$\boldsymbol{x}_{k+1} = \boldsymbol{A}\boldsymbol{x}_k + \boldsymbol{B}\boldsymbol{u}_k + \boldsymbol{L}_1\boldsymbol{f}_k, \tag{2.5}$$

$$\boldsymbol{y}_k = \boldsymbol{C}\boldsymbol{x}_k + \boldsymbol{D}\boldsymbol{u}_k + \boldsymbol{L}_2\boldsymbol{f}_k. \tag{2.6}$$

The redundancy relation can analytically be specified as follows: Combining together (2.5)–(2.6) from the time instant $k-s$ up to $k$ yields

$$\underbrace{\begin{bmatrix} \boldsymbol{y}_{k-S} \\ \boldsymbol{y}_{k-S+1} \\ \vdots \\ \boldsymbol{y}_k \end{bmatrix}}_{\boldsymbol{Y}_k} = \boldsymbol{H}\underbrace{\begin{bmatrix} \boldsymbol{u}_{k-S} \\ \boldsymbol{u}_{k-S+1} \\ \vdots \\ \boldsymbol{u}_k \end{bmatrix}}_{\boldsymbol{U}_k} = \boldsymbol{W}\boldsymbol{x}_{k-S} + \boldsymbol{M}\underbrace{\begin{bmatrix} \boldsymbol{f}_{k-S} \\ \boldsymbol{f}_{k-S+1} \\ \vdots \\ \boldsymbol{f}_k \end{bmatrix}}_{\boldsymbol{F}_k}, \tag{2.7}$$

whereas

$$\boldsymbol{H} = \begin{bmatrix} \boldsymbol{D} & \boldsymbol{0} & \cdots & \boldsymbol{0} \\ \boldsymbol{C}\boldsymbol{B} & \boldsymbol{D} & \cdots & \boldsymbol{0} \\ \vdots & \vdots & \ddots & \vdots \\ \boldsymbol{C}\boldsymbol{A}^{S-1}\boldsymbol{B} & \boldsymbol{C}\boldsymbol{A}^{S-1}\boldsymbol{B} & \cdots & \boldsymbol{D} \end{bmatrix}, \quad \boldsymbol{W} = \begin{bmatrix} \boldsymbol{C} \\ \boldsymbol{C}\boldsymbol{A} \\ \vdots \\ \boldsymbol{C}\boldsymbol{A}^S \end{bmatrix}, \tag{2.8}$$

and

$$\boldsymbol{M} = \begin{bmatrix} \boldsymbol{L}_2 & \boldsymbol{0} & \cdots & \boldsymbol{0} \\ \boldsymbol{C}\boldsymbol{L}_1 & \boldsymbol{L}_2 & \cdots & \boldsymbol{0} \\ \vdots & \vdots & \ddots & \vdots \\ \boldsymbol{C}\boldsymbol{A}^{S-1}\boldsymbol{L}_1 & \boldsymbol{C}\boldsymbol{A}^{S-1}\boldsymbol{L}_1 & \cdots & \boldsymbol{L}_2 \end{bmatrix}. \tag{2.9}$$

The residual signal can now be defined as [31]:

$$\boldsymbol{z}_k = \boldsymbol{V}[\boldsymbol{Y}_k - \boldsymbol{H}\boldsymbol{U}_k] = \boldsymbol{V}\boldsymbol{W}\boldsymbol{x}_{k-S} + \boldsymbol{V}\boldsymbol{M}\boldsymbol{F}_k. \tag{2.10}$$

In order to make (2.10) useful for fault detection, the matrix $\boldsymbol{V}$ should make the residual signal insensitive to system inputs and states, i.e., $\boldsymbol{V}\boldsymbol{W} = \boldsymbol{0}$. On the other hand, to make fault detection possible, the matrix $\boldsymbol{V}$ should also satisfy the condition $\boldsymbol{V}\boldsymbol{M} \neq \mathbf{0}$. It can be shown that for appropriately large $S$ (see [27] for how to obtain the minimum order $S$), it follows from the Cayley-Hamilton theorem that the solution to $\boldsymbol{V}\boldsymbol{W} = \mathbf{0}$ always exists. Finally, fault detection boils down to checking the norm of the residual, i.e., $\|\boldsymbol{z}_k\|$.

The fault isolation strategy can relatively easily be realised for sensor faults. Indeed, using the general idea of the dedicated fault isolation scheme, it is possible to design the parity relation with the $i$th, $i = 1, \ldots, m$, sensor only. This means that $y_i$ and $\boldsymbol{c}_i$ (being the $i$th row of $\boldsymbol{C}$) should be used instead of $\boldsymbol{y}$ and $\boldsymbol{C}$ in (2.6). Consequently, $\boldsymbol{D}$ should also be replaced by its $i$th row in (2.6). Thus, by assuming that all actuators all fault free, the $i$th residual generator is sensitive to the $i$th sensor fault only. This form of parity relation is called the single-sensor parity relation and it has been studied in a number of papers, e.g., [113, 141].

Unfortunately, the design strategy for actuator faults is not as straightforward as that for sensor faults. It can, of course, be realised in a very similar way but, as is indicated in [27, 113], the isolation of actuator faults is not always possible in the so-called single-actuator parity relation scheme.

### 2.1.3 Observers

The basic idea underlying observer-based (or filter-based, in the stochastic case) approaches to fault detection is to obtain the estimates of certain measured and/or unmeasured signals. Then, in the most usual case, the estimates of the measured signals are compared with their originals, i.e., the difference between the original signal and its estimate is used to form a residual signal $\boldsymbol{z}_k = \boldsymbol{y}_k - \hat{\boldsymbol{y}}_k$ (Fig. 2.2). To tackle this problem, many different observers (or filters) can be employed, e.g., Luenberger observers, Kalman filters, etc. From the above discussion, it is clear that the main objective is the estimation of system outputs while the estimation of the entire state vector is unnecessary. Since reduced-order observers can be employed, state estimation is significantly facilitated. On the other hand, to provide an additional freedom to achieve the required diagnostic performance, the observer order is usually larger than the possible minimum one.

The admiration for observer-based fault detection schemes is caused by the still increasing popularity of state-space models as well as the wide usage of observers in modern control theory and applications. Due to such conditions, the theory of observers (or filters) seems to be well developed (especially for linear systems). This has made a good background for the development of observer-based FDI schemes.

#### Luenberger observers and Kalman filters

Let us consider a linear system described by the following state-space equations:

$$\boldsymbol{x}_{k+1} = \boldsymbol{A}_k\boldsymbol{x}_k + \boldsymbol{B}_k\boldsymbol{u}_k + \boldsymbol{L}_{1,k}\boldsymbol{f}_k, \tag{2.11}$$

$$\boldsymbol{y}_{k+1} = \boldsymbol{C}_{k+1}\boldsymbol{x}_{k+1} + \boldsymbol{L}_{2,k+1}\boldsymbol{f}_{k+1}. \tag{2.12}$$

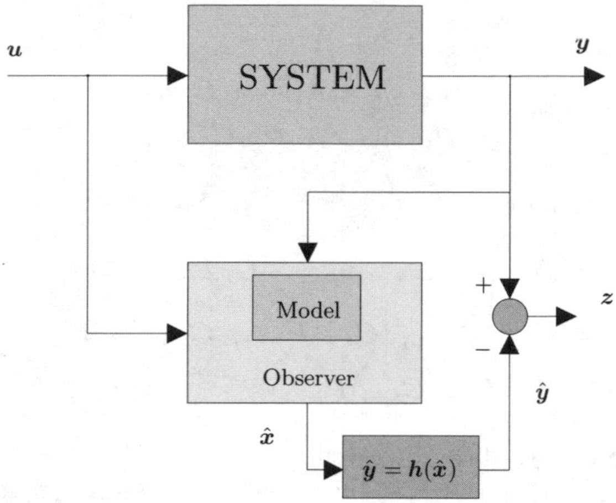


**Fig. 2.2.** Principle of observer-based residual generation

According to the observer-based residual generation scheme (Fig. 2.2), the residual signal can be given as

$$\begin{aligned}\boldsymbol{z}_{k+1} =& \boldsymbol{y}_{k+1} - \hat{\boldsymbol{y}}_{k+1} = \boldsymbol{C}_{k+1}[\boldsymbol{x}_{k+1} - \hat{\boldsymbol{x}}_{k+1}] + \boldsymbol{L}_{2,k+1}\boldsymbol{f}_{k+1} \\ =& \boldsymbol{C}_{k+1}[\boldsymbol{A}_k - \boldsymbol{K}_{k+1}\boldsymbol{C}_{k+1}][\boldsymbol{x}_k - \hat{\boldsymbol{x}}_k] + \boldsymbol{C}_{k+1}\boldsymbol{L}_{1,k}\boldsymbol{f}_k \\ &- \boldsymbol{C}_{k+1}\boldsymbol{K}_{k+1}\boldsymbol{L}_{2,k}\boldsymbol{f}_k + \boldsymbol{L}_{2,k+1}\boldsymbol{f}_{k+1}. \end{aligned} \tag{2.13}$$

To tackle the state estimation problem, the Luenberger observer can be used, i.e.,

$$\hat{\boldsymbol{x}}_{k+1} = \boldsymbol{A}_k\hat{\boldsymbol{x}}_k + \boldsymbol{B}_k\boldsymbol{u}_k + \boldsymbol{K}_{k+1}(\boldsymbol{y}_k - \hat{\boldsymbol{y}}_k), \tag{2.14}$$

and $\boldsymbol{K}_k$ stands for the so-called gain matrix and should be obtained in such a way as to ensure asymptotic convergence of the observer, i.e., $\lim_{k\to\infty}(\boldsymbol{x}_k - \hat{\boldsymbol{x}}_k) = \boldsymbol{0}$ [134]. If this is the case, i.e., $\hat{\boldsymbol{x}}_k \to \boldsymbol{x}_k$, the state estimation error $\boldsymbol{x}_k - \hat{\boldsymbol{x}}_k$ approaches zero and hence the residual signal (2.13) is only affected by the fault vector $\boldsymbol{f}_k$.

A similar approach can be realised in a stochastic setting, i.e., for systems which can be described by

$$\boldsymbol{x}_{k+1} = \boldsymbol{A}_k\boldsymbol{x}_k + \boldsymbol{B}_k\boldsymbol{u}_k + \boldsymbol{L}_{1,k}\boldsymbol{f}_k + \boldsymbol{w}_k, \tag{2.15}$$

$$\boldsymbol{y}_{k+1} = \boldsymbol{C}_{k+1}\boldsymbol{x}_{k+1} + \boldsymbol{L}_{2,k+1}\boldsymbol{f}_{k+1} + \boldsymbol{v}_{k+1}, \tag{2.16}$$

where $\boldsymbol{w}_k$ and $\boldsymbol{v}_k$ are zero-mean white noise sequences with the covariance matrices $\boldsymbol{Q}_k$ and $\boldsymbol{R}_k$, respectively. In this case, the observer structure can be similar to that of the Luenberger observer (2.14). To tackle the state estimation problem, the celebrated Kalman filter can be employed [4]. The algorithm of the Kalman filter can be described as follows:

1. Time update:

$$\hat{\boldsymbol{x}}_{k+1/k} = \boldsymbol{A}_k\hat{\boldsymbol{x}}_k + \boldsymbol{B}_k\boldsymbol{u}_k, \tag{2.17}$$

$$\boldsymbol{P}_{k+1/k} = \boldsymbol{A}_k\boldsymbol{P}_k\boldsymbol{A}_k^T + \boldsymbol{Q}_k. \tag{2.18}$$

2. Measurement update:

$$\hat{\boldsymbol{x}}_{k+1} = \hat{\boldsymbol{x}}_{k+1/k} + \boldsymbol{K}_{k+1}[\boldsymbol{y}_{k+1} - \boldsymbol{C}_{k+1}\hat{\boldsymbol{x}}_{k+1/k}], \tag{2.19}$$

$$\boldsymbol{K}_{k+1} = \boldsymbol{P}_{k+1/k}\boldsymbol{C}_{k+1}^T\left[\boldsymbol{C}_{k+1}\boldsymbol{P}_{k+1/k}\boldsymbol{C}_{k+1}^T + \boldsymbol{R}_{k+1}\right]^{-1}, \tag{2.20}$$

$$\boldsymbol{P}_{k+1} = [\boldsymbol{I} - \boldsymbol{K}_{k+1}\boldsymbol{C}_{k+1}]\boldsymbol{P}_{k+1/k}. \tag{2.21}$$

Finally, the residual signal can be given as

$$\begin{aligned}\boldsymbol{z}_{k+1} =& \boldsymbol{C}_{k+1}\boldsymbol{Z}_{k+1}\boldsymbol{A}_k[\boldsymbol{x}_k - \hat{\boldsymbol{x}}_k] + \boldsymbol{C}_{k+1}\boldsymbol{Z}_{k+1}\boldsymbol{L}_{1,k}\boldsymbol{f}_k \\ &+ \boldsymbol{M}_{k+1}\boldsymbol{L}_{2,k}\boldsymbol{f}_{k+1} + \boldsymbol{C}_{k+1}\boldsymbol{Z}_{k+1}\boldsymbol{w}_k + \boldsymbol{M}_{k+1}\boldsymbol{v}_{k+1},\end{aligned} \tag{2.22}$$

where $\boldsymbol{Z}_{k+1} = [\boldsymbol{I} - \boldsymbol{K}_{k+1}\boldsymbol{C}_{k+1}]$ and $\boldsymbol{M}_{k+1} = [\boldsymbol{I} - \boldsymbol{C}_{k+1}\boldsymbol{K}_{k+1}]$. Since the state estimate $\hat{\boldsymbol{x}}_k$ approaches the real state $\boldsymbol{x}_k$ (in the mean sense) asymptotically, i.e., $\mathcal{E}(\hat{\boldsymbol{x}}_k) \to \boldsymbol{x}_k$, the residual signal is only affected by faults and noise.

In both the deterministic (the Luenberger observer) and stochastic (the Kalman filter) cases, fault detection can be performed by checking that the residual norm $\|\boldsymbol{z}_k\|$ exceeds a prespecified threshold, i.e., $\|\boldsymbol{z}_k\| > T_H$. In the stochastic case, it is also possible to use more sophisticated, hypothesis-testing approaches such as Generalised Likelihood Ratio Testing (GLRT) or Sequential Probability Ratio Testing (SPRT) [13, 101, 178].

Observer-based fault isolation can, similarly to the parity relation approach described in Section 2.1.2, be implemented with the dedicated fault isolation scheme [84]. The main limitation of such an approach is related to the fact that by using a single output for sensor fault isolation the reduced system may not be observable. A similar limitation exists in the case of actuator fault isolation. A more efficient way of isolating faults is to use the generalised observer-based scheme [27]. Thus, by replacing the output vector $\boldsymbol{y}_k$ by $\boldsymbol{y}_k = [y_{1,k}, \ldots, y_{i-1}, y_{i+1}, \ldots, y_{m,k}]^T$ in (2.12) (or in (2.16)), it is possible to design an observer that is sensitive to all but the $i$th sensor fault. A similar strategy can be implemented for actuator faults with the so-called unknown input observers (see Section 2.3) or eigenstructure assignment [27]. However, similarly as is indicated in Section 2.1.2, the isolation of actuator faults is not always possible.

## 2.2 Approaches to Non-linear Systems

The main objective of this section is to present an introduction to the most popular FDI approaches for non-linear systems. Most techniques presented here constitute direct extensions of the approaches described in Section 2.1. Thus, FDI schemes based on these techniques are not described because they are the same as their linear counterparts. This means that the material of this section is reduced to the description of the heart of model-based FDI, which is a residual generator. There are, of course, some exceptions from this rule, but in such cases the FDI procedure is carefully described.

### 2.2.1 Parameter Estimation

Similarly as in the case of linear-in-parameter systems, the FDI problem boils down to estimating the parameters of the model of the system (Fig. 2.1). The system can be generally described by

$$y_k = g(\boldsymbol{\phi}_k, \boldsymbol{p}_k) + v_k, \tag{2.23}$$

while $\boldsymbol{\phi}_k$ may contain the previous or current system input $\boldsymbol{u}_k$, the previous system or model output, and the previous prediction error. The approach presented here inherits all the drawbacks and advantages of its linear counterpart presented in Section 2.1.1. The FDI scheme is also the same. Another problem arises because $g(\cdot)$ is non-linear in its parameters. In this case, non-linear parameter estimation techniques should be applied [170]. For complex models, this may cause serious difficulties with a fast reaction on faults; consequently, fault detection cannot be performed effectively and reliably. Irrespective of the above difficulties, there are, of course, some studies in which such an approach is utilised, e.g., [169].

### 2.2.2 Parity Relation

An extension of the parity relation to non-linear polynomial dynamic systems was proposed in [70]. In order to describe this approach, let us consider a system described by the state-space equations

$$\boldsymbol{x}_{k+1} = \boldsymbol{g}\left(\boldsymbol{x}_k, \boldsymbol{u}_k, \boldsymbol{f}_k\right), \tag{2.24}$$

$$\boldsymbol{y}_k = \boldsymbol{h}(\boldsymbol{x}_k, \boldsymbol{u}_k, \boldsymbol{f}_k), \tag{2.25}$$

where $\boldsymbol{g}(\cdot)$ and $\boldsymbol{h}(\cdot)$ are assumed to be polynomials. The equations (2.24)–(2.25) can always be expressed on a time window $[k-S, k]$. As a result, the following structure can be obtained:

$$\boldsymbol{y}_{k-S,k} = \boldsymbol{H}\left(\boldsymbol{x}_{k-S}, \boldsymbol{u}_{k-S,k}, \boldsymbol{f}_{k-S,k}\right), \tag{2.26}$$

with $\boldsymbol{u}_{k-S,k} = \boldsymbol{u}_{k-S}, \ldots, \boldsymbol{u}_k$ and $\boldsymbol{f}_{k-S,k} = \boldsymbol{f}_{k-S}, \ldots, \boldsymbol{f}_k$. In order to check the consistency of the model equations, the state variables have to be eliminated. This results in the following equation:

$$\boldsymbol{\Phi}(\boldsymbol{y}_{k-S,k}, \boldsymbol{u}_{k-S,k}, \boldsymbol{f}_{k-S,k}) = \boldsymbol{0}. \tag{2.27}$$

Since $\boldsymbol{g}(\cdot)$ and $\boldsymbol{h}(\cdot)$ are assumed to be polynomials, elimination theory can be applied to transform (2.26) into (2.27). Knowing that the $\Phi_i(\cdot)$s are polynomials and therefore they are expressed as sums of monomials, it seems natural to split the expression (2.27) into two parts, i.e.,

$$\boldsymbol{z}_k = \boldsymbol{\Phi}_1(\boldsymbol{y}_{k-S,k}, \boldsymbol{u}_{k-S,k}), \tag{2.28}$$

$$\boldsymbol{z}_k = \boldsymbol{\Phi}_2(\boldsymbol{y}_{k-S,k}, \boldsymbol{u}_{k-S,k}, \boldsymbol{f}_{k-S,k}). \tag{2.29}$$

The right-hand side of (2.28) contains all the monomials in $\boldsymbol{y}_{k-S,k}$ and $\boldsymbol{u}_{k-S,k}$ only, while (2.29) contains all the monomials involving at least one of the components of $\boldsymbol{f}_{k-S,k}$. The above condition ensures that $\boldsymbol{z}_k = \boldsymbol{0}$ in the fault-free case. Since the fault signal $\boldsymbol{f}_{k-S,k}$ is not measurable, only the equation (2.28) can be applied to generate the residual $\boldsymbol{z}_k$ and, consequently, to detect faults.

One drawback of this approach is that it is limited to polynomial models or, more precisely, to models for which the state vector $\boldsymbol{x}_k$ can be eliminated. Another drawback is that it is assumed that a perfect model is available, i.e., there is no model uncertainty. This may cause serious problems while applying the approach to real systems.

Parity relation for a more general class of non-linear systems was proposed by Krishnaswami and Rizzoni [104]. The FDI scheme considered is shown in Fig. 2.3. There are two residual vectors, namely, the forward $\boldsymbol{z}_{f,k}$ residual vector and the backward $\boldsymbol{z}_{b,k}$ residual vector. These residuals are generated using the forward and inverse (backward) models, respectively. Based on these residual vectors, fault detection can (theoretically) be easily performed, while fault isolation should be realised according to Tab. 2.1. The authors suggest an extension of the proposed approach to cases where model uncertainty is considered.

**Table 2.1.** Principle of fault isolation with the non-linear parity relation

| Fault location | Non-zero element of $\boldsymbol{z}_{f,k}$ | Non-zero element of $\boldsymbol{z}_{b,k}$ |
|---|---|---|
| $i$th sensor | $z_f^i$ | all elements dependent on $y_i$ |
| $i$th actuator | all elements dependent on $u_i$ | $z_b^i$ |

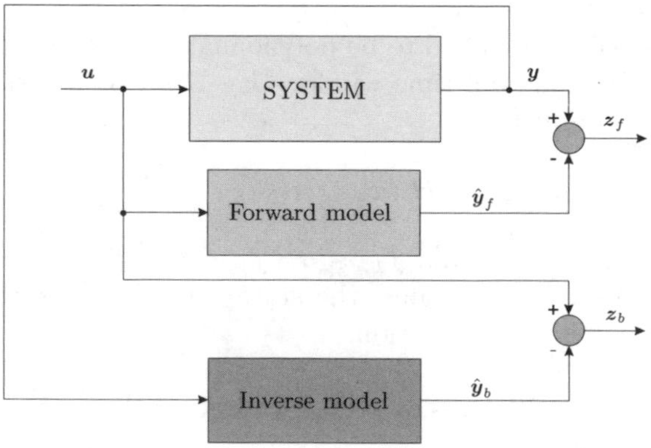


**Fig. 2.3.** Non-linear parity relation-based FDI

Undoubtedly, strict existence conditions for an inverted model as well as possible difficulties with the application of the known identification techniques make the usefulness of this approach for a wide class of non-linear systems questionable.

Another parity relation approach for non-linear systems was proposed by Shumsky [159]. The concepts of the parity relation and parameter estimation fault detection techniques are combined. In particular, the parity relation is used to detect offsets in the model parameters. The necessary condition is that there exists a transformation $\boldsymbol{x}_k = \boldsymbol{\xi}(\boldsymbol{u}_k, \ldots, \boldsymbol{u}_{k+S}, \boldsymbol{y}_k, \ldots, \boldsymbol{y}_{k+S})$, which may cause serious problems in many practical applications. Another inconvenience is that the approach inherits most drawbacks concerning parameter estimation-based fault detection techniques.

### 2.2.3 Observers

Model linearisation is a straightforward way of extending the applicability of linear techniques to non-linear systems. On the other hand, it is well known that such approaches work well when there is no large mismatch between the linearised model and the non-linear system. Two types of linearisation can be distinguished, i.e., linearisation around the constant state and linearisation around the current state estimate. It is obvious that the second type of linearisation usually yields better results. Unfortunately, during such a linearisation the influence of terms higher than linear is usually neglected (as in the case of the extended Luenberger observer and the extended Kalman filter). One way out from this problem is to improve the performance of linearisation-based observers. Another way is to use linearisation-free approaches. Unfortunately, the application of such observers is limited to certain classes of non-linear systems.

Generally, FDI principles of non-linear observer-based schemes are not different than those described in Section 2.1.3. Apart from this similarity, the design complexity and feasibility of such FDI schemes is usually far more sophisticated.

#### Extended Luenberger observers and Kalman filters

Let us consider a non-linear discrete-time system described by the following state-space equations:

$$\boldsymbol{x}_{k+1} = \boldsymbol{g}\left(\boldsymbol{x}_k, \boldsymbol{u}_k\right) + \boldsymbol{L}_{1,k}\boldsymbol{f}_k, \tag{2.30}$$

$$\boldsymbol{y}_{k+1} = \boldsymbol{h}(\boldsymbol{x}_{k+1}) + \boldsymbol{L}_{2,k+1}\boldsymbol{f}_{k+1}. \tag{2.31}$$

In order to apply the Luenberger observer presented in Section 2.1.3, it is necessary to linearise the equations (2.30) and (2.31) around either a constant value (e.g., $\boldsymbol{x} = \boldsymbol{0}$) or the current state estimate $\hat{\boldsymbol{x}}_k$. The latter approach seems to be more appropriate as it improves its approximation accuracy as $\hat{\boldsymbol{x}}_k$ tends to $\boldsymbol{x}_k$. In this case, the approximation can be realised as follows:

$$\boldsymbol{A}_k = \left.\frac{\partial \boldsymbol{g}\left(\boldsymbol{x}_k, \boldsymbol{u}_k\right)}{\partial \boldsymbol{x}_k}\right|_{\boldsymbol{x}_k = \hat{\boldsymbol{x}}_k}, \quad \boldsymbol{C}_k = \left.\frac{\partial \boldsymbol{h}(\boldsymbol{x}_k)}{\partial \boldsymbol{x}_k}\right|_{\boldsymbol{x}_k = \hat{\boldsymbol{x}}_k}. \tag{2.32}$$

As a result of using the Luenberger observer (2.14), the state estimation error takes the form

$$\begin{aligned}\boldsymbol{e}_{k+1} = [\boldsymbol{A}_{k+1} - \boldsymbol{K}_{k+1}\boldsymbol{C}_k]\boldsymbol{e}_k + \boldsymbol{L}_{1,k}\boldsymbol{f}_k - \boldsymbol{K}_{k+1}\boldsymbol{L}_{2,k}\boldsymbol{f}_k + \\ + \boldsymbol{o}(\boldsymbol{x}_k, \hat{\boldsymbol{x}}_k),\end{aligned} \tag{2.33}$$

where $\boldsymbol{o}(\boldsymbol{x}_k, \hat{\boldsymbol{x}}_k)$ stands for the linearisation error caused by the introduction of (2.32).

Because of a highly time-varying nature of $\boldsymbol{A}_k$ and $\boldsymbol{C}_k$ as well as the linearisation error $\boldsymbol{o}(\boldsymbol{x}_k, \hat{\boldsymbol{x}}_k)$, it is usually very difficult to obtain an appropriate form of the gain matrix $\boldsymbol{K}_{k+1}$. This is the main reason why this approach is rarely used in practice.

As the Kalman filter constitutes a stochastic counterpart of the Luenberger observer, the extended Kalman filter can also be designed for the following class of non-linear systems:

$$\boldsymbol{x}_{k+1} = \boldsymbol{g}(\boldsymbol{x}_k, \boldsymbol{u}_k) + \boldsymbol{L}_{1,k}\boldsymbol{f}_k + \boldsymbol{w}_k, \tag{2.34}$$

$$\boldsymbol{y}_{k+1} = \boldsymbol{h}(\boldsymbol{x}_{k+1}) + \boldsymbol{L}_{2,k+1}\boldsymbol{f}_{k+1} + \boldsymbol{v}_{k+1}, \tag{2.35}$$

while, similarly to the linear case, $\boldsymbol{w}_k$ and $\boldsymbol{v}_k$ are zero-mean white noise sequences. Using the linearisation (2.32) and neglecting the influence of the linearisation error, it is straightforward to use the Kalman filter algorithm described in Section 2.1.3. The main drawback of such an approach is that it works well only when there is no large mismatch between the model linearised around the current state estimate and the non-linear behaviour of the system.

The EKF can also be used for deterministic systems, i.e., as an observer for the system (2.30)–(2.31) (see [19] and the references therein). In this case, the noise covariance matrices can be set almost arbitrarily. As was proposed in [19], this possibility can be used to increase the convergence of an observer.

Apart from difficulties regarding linearisation errors, similarly to the case of linear systems, the presented approaches do not take model uncertainty into account. This drawback disqualifies those techniques for most practical applications, although there are cases for which such techniques work with acceptable efficiency, e.g., [100].

### Observers for Lipschitz systems

Let us consider a class of non-linear systems which can be described by the following state-space equations:

$$\boldsymbol{x}_{k+1} = \boldsymbol{A}\boldsymbol{x}_k + \boldsymbol{B}\boldsymbol{u}_k + \boldsymbol{h}(\boldsymbol{y}_k, \boldsymbol{u}_k) + \boldsymbol{g}(\boldsymbol{x}_k, \boldsymbol{u}_k), \tag{2.36}$$

$$\boldsymbol{y}_{k+1} = \boldsymbol{C}\boldsymbol{x}_{k+1}, \tag{2.37}$$

and $\boldsymbol{g}(\boldsymbol{x}_k, \boldsymbol{u}_k)$ satisfies

$$\|\boldsymbol{g}(\boldsymbol{x}_1, \boldsymbol{u}) - \boldsymbol{g}(\boldsymbol{x}_2, \boldsymbol{u})\|_2 \leq \gamma\|\boldsymbol{x}_1 - \boldsymbol{x}_2\|_2, \; \forall \boldsymbol{x}_1, \boldsymbol{x}_2, \boldsymbol{u}, \tag{2.38}$$

where $\gamma > 0$ stands for the Lipschitz constant. Many non-linear systems can be described by (2.36), e.g., sinusoidal non-linearities satisfy (2.38), even polynomial non-linearities satisfy (2.38) assuming that $\boldsymbol{x}_k$ is bounded. This means that (2.36)–(2.37) can be used for describing a wide class of non-linear systems, which is very important from the point of view of potential industrial applications.

The first solution for state estimation of the continuous-time counterpart of (2.36)–(2.37) was developed by Thau [166]. Assuming that the pair $(\boldsymbol{A}, \boldsymbol{C})$ is observable, Thau proposed a convergence condition, but he did not provide an effective design procedure for the observer. In other words, in light of this approach, the observer has to be designed with a trial-and-error procedure that amounts to solving a large number of Lyapunov equations and then checking the convergence conditions. Many different authors followed a similar procedure but they proposed less restrictive convergence conditions (see, e.g., [155]). Finally, in [1, 149, 150] the authors proposed a more effective observer design. In particular, in [150] the authors employed the concept of the distance to the unobservability of the pair $(\boldsymbol{A}, \boldsymbol{C})$, and proposed an iterative coordinate transformation technique reducing the Lipschitz constant. In [1] the authors employed and improved the results of [150], but the proposed design procedure does not seem straightforward. In [149], the author reduced the observer design problem to a global optimisation one. The main disadvantage of this approach is that the proposed algorithm does not guarantee obtaining a global optimum. Thus, many trial-and-error steps have to be carried out to obtain a satisfactory solution. Recently, in [142] the authors proposed the so-called dynamic observer with a mixed binary search and the $\mathcal{H}_\infty$ optimisation procedure.

Unfortunately, the theory and practice concerning observers for discrete-time systems (2.36)–(2.37) are significantly less mature than those for their continuous-time counterparts. Indeed, there are few papers only [22, 172] dealing with discrete-time observers. The authors of the above works proposed different parameterisations of the observer, but the common disadvantage of these approaches is that a trial-and-error procedure has to be employed that boils down to solving a large number of Lyapunov equations. Moreover, the authors do not provide convergence conditions similar to those for continuous-time observers [155, 166].

### Coordinate change-based observers

Another possible approach can be implemented by a suitable non-linear change of coordinates to bring the original system into a linear (or pseudo-linear) one. Let us consider the following non-linear system:

$$\boldsymbol{x}_{k+1} = \boldsymbol{g}\left(\boldsymbol{x}_k, \boldsymbol{u}_k\right) + \boldsymbol{L}_{1,k}\boldsymbol{f}_k, \tag{2.39}$$

$$\boldsymbol{y}_{k+1} = \boldsymbol{h}(\boldsymbol{x}_{k+1}) + \boldsymbol{L}_{2,k+1}\boldsymbol{f}_{k+1}. \tag{2.40}$$

For design purposes, let us assume that $\boldsymbol{f}_k = \boldsymbol{0}$. The basic idea underlying coordinate-change-based observers is to determine the coordinate change (at least locally defined) of the form

$$\begin{aligned} \boldsymbol{s} &= \boldsymbol{\phi}(\boldsymbol{x}), \\ \bar{\boldsymbol{y}} &= \boldsymbol{\varphi}(\boldsymbol{y}), \end{aligned} \tag{2.41}$$

such that in the new coordinates (2.41) the system (2.39)–(2.40) is described by:

$$\boldsymbol{s}_{k+1} = \boldsymbol{A}(\boldsymbol{u}_k)\boldsymbol{s}_k + \boldsymbol{\psi}(\boldsymbol{y}_k, \boldsymbol{u}_k), \tag{2.42}$$

$$\bar{\boldsymbol{y}}_{k+1} = \boldsymbol{C}\boldsymbol{s}_{k+1}, \tag{2.43}$$

where $\boldsymbol{\psi}(\cdot)$ is a non-linear (in general) function. There is no doubt that the observer design problem is significantly simplified when (2.42)–(2.43) are used instead of (2.39)–(2.40). The main drawback of such an approach is related to strong design conditions that limit its application to particular classes of non-linear systems (for an example regarding single-input single-output systems, the reader is referred to [23]).

### Observers for bilinear and low-order polynomial systems

A polynomial (and, as a special case, bilinear) system description is a natural extension of linear models. Designs of observers for bilinear and low-order polynomial systems have been considered in a number of papers [6, 75, 81, 92, 158, 191]. Let us consider a bilinear system modelled by the following state space equations:

$$\boldsymbol{x}_{k+1} = \boldsymbol{A}_k\boldsymbol{x}_k + \sum_{i=1}^{r} \boldsymbol{A}^i u_{i,k}\boldsymbol{x}_k + \boldsymbol{B}\boldsymbol{u}_k + \boldsymbol{L}_1\boldsymbol{f}_k, \tag{2.44}$$

$$\boldsymbol{y}_{k+1} = \boldsymbol{C}\boldsymbol{x}_{k+1} + \boldsymbol{D}\boldsymbol{u}_k + \boldsymbol{L}_2\boldsymbol{f}_{k+1}. \tag{2.45}$$

Hou and Pugh [81] established the necessary conditions for the existence of the observer for the continuous-time counterpart of (2.44)–(2.45). Moreover, they proposed a design procedure involving the transformation of the original system (2.44)–(2.45) into an equivalent, quasi-linear one. An observer for systems which can be described by state-space equations consisting of both linear and polynomial terms was proposed in [6, 158].

## 2.3 Robustness Issues

Irrespective of the linear (Section 2.1) or non-linear (Section 2.2) FDI technique being employed, FDI performance will be usually impaired by the lack of robustness to model uncertainty. Indeed, the model reality mismatch may cause very undesirable situations such as undetected faults or false alarms. This may lead to serious economical losses or even catastrophes.

Taking into account the above conditions, a large amount of knowledge on designing robust fault diagnosis systems has been accumulated through the literature since the beginning of the 1980s. For a comprehensive survey regarding such techniques, the reader is referred to the excellent monographs [27, 65, 96, 138]. Thus, the subject of this section is to outline the main issues of robust fault diagnosis.

Let us start with the parameter estimation techniques described in Sections 2.1.1 and 2.2.1. The main assumption underlying the FDI approaches presented in these sections, was the fact that perfect parameter estimates can be obtained.

This is, of course, very hard to attain in practice. Indeed, the fact that the measurements used for parameter estimation can be corrupted by noise and disturbances contributes directly to the so-called parameter uncertainty. This means that there is no unique $\hat{\boldsymbol{p}}$ that is consistent with a given set of measurements, but there is a parameter set $\mathbb{P}$ that satisfies this requirement. This parameter set is usually called the confidence region or the feasible parameter set [116, 170]. Such a set can be determined on-line for linear-in-parameter systems using either statistical [170] or bounded-error approaches [116, 170, 173]. In both the cases, fault diagnosis tasks can be realised in two different ways. The first one boils down to

$$\boldsymbol{p}_{\text{nom}} \notin \mathbb{P}_k \quad \Rightarrow \quad \boldsymbol{f}_k \neq \mathbf{0}, \tag{2.46}$$

where $\mathbb{P}_k$ is the parameter confidence region associated with $k$ input-output measurements. The second approach is implemented in such a way as the knowledge regarding $\mathbb{P}_k$ is used to calculate the so-called output confidence interval:

$$y_k^N \leq y_k \leq y_k^M. \tag{2.47}$$

This confidence interval is then used for fault detection, i.e.,

$$y_k < y_k^N \quad \text{or} \quad y_k > y_k^M \quad \Rightarrow \quad \boldsymbol{f}_k \neq \mathbf{0}. \tag{2.48}$$

The advantage of (2.48) over (2.46) is that no knowledge regarding the nominal (fault-free) parameter vector $\boldsymbol{p}_{\text{nom}}$ is required.

In the case of fault isolation, the feasible parameter set $\mathbb{P}$ can be used for calculating the parameter confidence intervals:

$$p_{i,k}^N \leq p_i \leq p_{i,k}^M, \quad i = 1, \ldots, n_p. \tag{2.49}$$

These intervals can then be employed for fault isolation:

$$p_{i,\text{nom}} < p_{i,k}^N \quad \text{or} \quad p_{i,\text{nom}} > p_{i,k}^M \quad \Rightarrow \quad f_{i,k} \neq \mathbf{0}, \quad i = 1, \ldots, s = n_p. \tag{2.50}$$

In order to extend the above strategies for non-linear-in-parameter systems, it is necessary to use various linearisation strategies, which usually impair the reliability and effectiveness of FDI.

An alternative approach to robust parameter estimation-based fault diagnosis was proposed in [135, 193]. The authors considered the continuous-time counterpart of the following system:

$$\boldsymbol{x}_{k+1} = \boldsymbol{g}(\boldsymbol{x}_k, \boldsymbol{u}_k) + \boldsymbol{h}(\boldsymbol{x}_k, \boldsymbol{u}_k, k) + \boldsymbol{L}(k - k_0)\boldsymbol{\phi}(\boldsymbol{x}_k, \boldsymbol{u}_k), \tag{2.51}$$

while $\boldsymbol{L}(k - k_0)$ is a matrix function representing the time profiles of the faults (with $k_0$ being an unknown time), and $\boldsymbol{h}(\boldsymbol{x}_k, \boldsymbol{u}_k, k)$ is the modelling error satisfying

$$|h_i(\boldsymbol{x}_k, \boldsymbol{u}_k, k)| \leq \bar{h}, \quad i = 1, \ldots, n, \tag{2.52}$$

where $\bar{h}$ is a known bound. A change in the system behaviour caused by the faults is modelled by $\boldsymbol{\phi}(\boldsymbol{x}_k, \boldsymbol{u}_k)$, which belongs to a finite set of functions:

$$\mathbb{F} = \left\{\boldsymbol{\phi}^1(\boldsymbol{x}_k, \boldsymbol{u}_k), \ldots, \boldsymbol{\phi}^s(\boldsymbol{x}_k, \boldsymbol{u}_k)\right\}. \tag{2.53}$$

Each fault function $\boldsymbol{\phi}^i(\boldsymbol{x}_k, \boldsymbol{u}_k)$, $i = 1, \ldots, s$ is assumed to be a parametric fault, i.e., a fault with a known non-linear structure but with unknown parameter vectors. To tackle parametric fault detection, the authors proposed the so-called detection and approximation observer while fault isolation was realised with a bank of non-linear adaptive observers.

As can be observed in the literature [27, 65, 96, 138], the most common approach to robust fault diagnosis is to use robust observers. This is mainly because of the fact that the theory of robust observers is relatively well developed in the control engineering literature. Indeed, the most common approaches to representing model uncertainty in robust observers for linear systems can be divided into five categories:

Norm-bounded model uncertainty: it corresponds to a system description whose matrices are modelled in the form of a known matrix $\boldsymbol{M}_0$ and an additive uncertainty term $\boldsymbol{\Delta}_M$ satisfying $\|\boldsymbol{\Delta}_M\| \leq 1$. Thus the matrices describing the system are of the following form:

$$\boldsymbol{M} = \boldsymbol{M}_0 + \boldsymbol{\Delta}_M = \boldsymbol{M}_0 + \boldsymbol{HFE}, \tag{2.54}$$

where $\boldsymbol{F}$ and $\boldsymbol{E}$ are known constant matrices, and

$$\boldsymbol{F}^T\boldsymbol{F} \preceq \boldsymbol{I}. \tag{2.55}$$

Polytopic model uncertainty: it corresponds to a system description whose matrices are contained in a polytope of matrices, i.e.,

$$\boldsymbol{M} \in \mathrm{Co}\,(\boldsymbol{M}_{\mathbf{1}}, \ldots, \boldsymbol{M}_{\boldsymbol{N}}), \tag{2.56}$$

or, equivalently, $\boldsymbol{M} \in \mathbb{M}$ with

$$\mathbb{M} = \left\{\boldsymbol{X} : \boldsymbol{X} = \sum_{i=1}^{N} \alpha_i \boldsymbol{M}_i,\ \alpha_i \geq 0,\ \sum_{i=1}^{N} \alpha_i = 1\right\}. \tag{2.57}$$

Affine model uncertainty: it corresponds to a system description whose matrices are modelled as a collection of fixed affine functions of varying parameters $p_1, \ldots, p_{n_p}$, i.e.,

$$\boldsymbol{M}(\boldsymbol{p}) = \boldsymbol{M}_0 + p_1\boldsymbol{M}_1 + \cdots + p_{n_p}\boldsymbol{M}_{n_p}, \tag{2.58}$$

whereas $\boldsymbol{M}_i$, $i = 0, \ldots, n_p$ are known matrices, and

$$\underline{p}_i \leq p_i \leq \bar{p}_i, \quad i = 1, \ldots, n_p. \tag{2.59}$$

Interval model uncertainty: it corresponds to a system description whose matrices $\boldsymbol{M}$ are not known precisely but their elements are described by known intervals, i.e.:

$$\underline{m}_{i,j} \leq m_{i,j} \leq \bar{m}_{i,j}. \tag{2.60}$$

Unknown input model uncertainty: it corresponds to the system description in which model uncertainty is modelled by an unknown additive term, i.e.,

$$\boldsymbol{x}_{k+1} = \boldsymbol{A}\boldsymbol{x}_k + \boldsymbol{B}\boldsymbol{u}_k + \boldsymbol{E}\boldsymbol{d}_k, \tag{2.61}$$

where $\boldsymbol{d}_k$ is an unknown input, and $\boldsymbol{E}$ denotes its distribution matrix which is known, i.e., it can be efficiently estimated with one of the approaches described in [27, 96].

As can be found in the literature [27, 65, 84, 96, 138], the most popular approach is to use unknown input model uncertainty. The observer resulting from such an approach is called the Unknown Input Observer (UIO).

Although the origins of UIOs can be traced back to the early 1970s (cf. the seminal work of Wang *et al.* [171]), the problem of designing such observers is still of paramount importance both from the theoretical and practical viewpoints. A large amount of knowledge on using these techniques for model-based fault diagnosis has been accumulated through the literature for the last three decades (see [27, 96] and the references therein). Generally, design problems regarding UIOs can be divided into three distinct categories:

Design of UIOs for linear deterministic systems:
Apart from the seminal paper of Wang *et al.* [171], it is worth noting a few pioneering and important works in this area, namely, the geometric approach by Bhattacharyya [14], the inversion algorithm by Kobayashi and Nakamizo [93], the algebraic approach by Hou and Müller [79] and, finally, the approach by Chen, Patton and Zhang [29]. The reader is also referred to the recently published developments, e.g., [82].

Design of UIOs for linear stochastic systems:
Most design techniques concerning such a class of linear systems make use of ideas for linear deterministic systems along with the Kalman filtering strategy. Thus, the resulting approaches can be perceived as Kalman filters for linear systems with unknown inputs. The representative approaches of this group were developed by Chen, Patton and Zhang [27, 29], Darouach and Zasadzinski [37], Hou and Patton [80] and, finally, Keller and Darouach [91].

A significantly different approach was proposed in [179]. Instead of using the Kalman filter-like approach, the author employed the bounded-error state estimation technique [109], but the way of decoupling the unknown input remained the same as that in [91].

Design of UIOs for non-linear systems:
The design approaches developed for non-linear systems can generally be divided into three categories:

non-linear state-transformation-based techniques: apart from a relatively large class of systems to which they can be applied, even if the non-linear transformation is possible it sometimes leads to another non-linear system and hence the observer design problem remains open (see [2, 157] and the references therein).

linearisation-based techniques: such approaches are based on a similar strategy as that for the extended Kalman filter [96]. In [179, 187], the author proposed an Extended Unknown Input Observer (EUIO) for non-linear systems. He also proved that the proposed observer is convergent under certain conditions.

Observers for particular classes of non-linear systems: for example, UIOs for polynomial and bilinear systems [6, 75] or UIOs for Lipschitz systems [94, 142].

To illustrate the general principle of the UIO, let us consider a linear system described by the following state-space equations:

$$\boldsymbol{x}_{k+1} = \boldsymbol{A}_k\boldsymbol{x}_k + \boldsymbol{B}_k\boldsymbol{u}_k + \boldsymbol{E}_k\boldsymbol{d}_k + \boldsymbol{L}_{1,k}\boldsymbol{f}_k, \tag{2.62}$$

$$\boldsymbol{y}_{k+1} = \boldsymbol{C}_{k+1}\boldsymbol{x}_{k+1} + \boldsymbol{L}_{2,k+1}\boldsymbol{f}_{k+1}, \tag{2.63}$$

where the term $\boldsymbol{E}_k\boldsymbol{d}_k$ stands for model uncertainty as well as real disturbances acting on the system, and $\text{rank}(\boldsymbol{E}_k) = q$. The general structure of an unknown input observer can be given as follows [27]:

$$\boldsymbol{s}_{k+1} = \boldsymbol{F}_{k+1}\boldsymbol{s}_k + \boldsymbol{T}_{k+1}\boldsymbol{B}_k\boldsymbol{u}_k + \boldsymbol{K}_{k+1}\boldsymbol{y}_k, \tag{2.64}$$

$$\hat{\boldsymbol{x}}_{k+1} = \boldsymbol{s}_{k+1} + \boldsymbol{H}_{k+1}. \tag{2.65}$$

If the following relations hold true:

$$\boldsymbol{K}_{k+1} = \boldsymbol{K}_{1,k+1} + \boldsymbol{K}_{2,k+2}, \tag{2.66}$$

$$\boldsymbol{T}_{k+1} = \boldsymbol{I} - \boldsymbol{H}_{k+1}\boldsymbol{C}_{k+1} \tag{2.67}$$

$$\boldsymbol{F}_{k+1} = \boldsymbol{A}_k - \boldsymbol{H}_{k+1}\boldsymbol{C}_{k+1}\boldsymbol{A}_k - \boldsymbol{K}_{1,k+1}\boldsymbol{C}_k, \tag{2.68}$$

$$\boldsymbol{K}_{2,k+1} = \boldsymbol{F}_{k+1}\boldsymbol{H}_k, \tag{2.69}$$

then (assuming the fault-free mode, i.e., $\boldsymbol{f}_k = \boldsymbol{0}$ ) the state estimation error is

$$\boldsymbol{e}_{k+1} = \boldsymbol{F}_{k+1}\boldsymbol{e}_k + [\boldsymbol{I} - \boldsymbol{H}_{k+1}\boldsymbol{C}_{k+1}]\boldsymbol{E}_k\boldsymbol{d}_k. \tag{2.70}$$

From the above equation, it is clear that to decouple the effect of an unknown input from the state estimation error (and, consequently, from the residual), the following relation should be satisfied:

$$[\boldsymbol{I} - \boldsymbol{H}_{k+1}\boldsymbol{C}_{k+1}]\boldsymbol{E}_k = \boldsymbol{0}. \tag{2.71}$$

The necessary condition for the existence of a solution to (2.71) is $\text{rank}(\boldsymbol{E}_k) = \text{rank}(\boldsymbol{C}_{k+1}\boldsymbol{E}_k)$ [27, p. 72, Lemma 3.1], and a special solution is

$$\boldsymbol{H}^*_{k+1} = \boldsymbol{E}_k\left[(\boldsymbol{C}_{k+1}\boldsymbol{E}_k)^T\boldsymbol{C}_{k+1}\boldsymbol{E}_k\right]^{-1}(\boldsymbol{C}_{k+1}\boldsymbol{E}_k)^T. \tag{2.72}$$

The remaining task is to design the matrix $\boldsymbol{K}_{1,k+1}$ so as to ensure the convergence of the observer. This can be realised in a similar way as it is done in the case of the Luenberger observer. Finally, the state estimation error and the residual are given by

$$\begin{aligned}\boldsymbol{e}_{k+1} &= \boldsymbol{F}_{k+1}\boldsymbol{e}_k + \boldsymbol{T}_{k+1}\boldsymbol{L}_{1,k}\boldsymbol{f}_k \\ &\quad - \boldsymbol{H}_{k+1}\boldsymbol{L}_{2,k+1}\boldsymbol{f}_{k+1} - \boldsymbol{K}_{1,k+1}\boldsymbol{L}_{2,k}\boldsymbol{f}_k, \quad (2.73)\\ \boldsymbol{z}_{k+1} &= \boldsymbol{C}_{k+1}\boldsymbol{e}_{k+1} + \boldsymbol{L}_{2,k+1}\boldsymbol{f}_{k+1}. \quad (2.74)\end{aligned}$$

Since the Kalman filter constitutes a stochastic counterpart of the Luenberger observer, there can also be developed a stochastic counterpart of the UIO [27], i.e., an observer which can be applied to the following class of systems:

$$\begin{aligned}\boldsymbol{x}_{k+1} &= \boldsymbol{A}_k\boldsymbol{x}_k + \boldsymbol{B}_k\boldsymbol{u}_k + \boldsymbol{E}_k\boldsymbol{d}_k + \boldsymbol{L}_{1,k}\boldsymbol{f}_k + \boldsymbol{w}_k, \quad (2.75)\\ \boldsymbol{y}_{k+1} &= \boldsymbol{C}_k\boldsymbol{x}_{k+1} + \boldsymbol{L}_{2,k+1}\boldsymbol{f}_{k+1} + \boldsymbol{v}_{k+1}. \quad (2.76)\end{aligned}$$

Apart from the robustness properties, another reason why UIOs are very popular in fault diagnosis schemes is the fact that they can be effectively applied to sensor and actuator fault isolation. First, the sensor fault isolation scheme is described. In this case, the actuators are assumed to be fault-free, and hence, for each of the observers, the system can be characterised as follows:

$$\begin{aligned}\boldsymbol{x}_{k+1} &= \boldsymbol{g}(\boldsymbol{x}_k) + \boldsymbol{h}(\boldsymbol{u}_k) + \boldsymbol{E}_k\boldsymbol{d}_k, \quad (2.77)\\ \boldsymbol{y}^j_{k+1} &= \boldsymbol{C}^j_{k+1}\boldsymbol{x}_{k+1} + \boldsymbol{f}^j_{k+1}, \qquad j = 1,\ldots,m, \quad (2.78)\\ y_{j,k+1} &= \boldsymbol{c}_{j,k+1}\boldsymbol{x}_{k+1} + f_{j,k+1}, \quad (2.79)\end{aligned}$$

where, similarly as it was done in [27], $\boldsymbol{c}_{j,k} \in \mathbb{R}^n$ is the $j$th row of the matrix $\boldsymbol{C}_k$, $\boldsymbol{C}^j_k \in \mathbb{R}^{m-1\times n}$ is obtained from the matrix $\boldsymbol{C}_k$ by deleting the $j$th row, $\boldsymbol{c}_{j,k}$, $y_{j,k+1}$ is the $j$th element of $\boldsymbol{y}_{k+1}$, and $\boldsymbol{y}^j_{k+1} \in \mathbb{R}^{m-1}$ is obtained from the vector $\boldsymbol{y}_{k+1}$ by deleting the $j$th component $y_{j,k+1}$. Thus, the problem reduces to designing $m$ UIOs (Fig. 2.4). Therefore, each residual generator (observer) is driven by all inputs and all outputs but one. When all sensors but the $j$th one

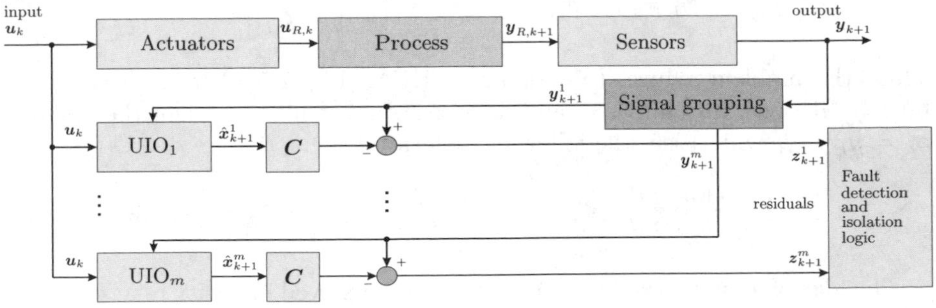


**Fig. 2.4.** Sensor fault detection and isolation scheme

are fault-free and all actuators are fault-free, then the residual $\boldsymbol{z}_k = \boldsymbol{y}_k - \hat{\boldsymbol{y}}_k$ will satisfy the following isolation logic:

$$\begin{cases} \|\boldsymbol{z}_k^j\| < T_j^H \\ \|\boldsymbol{z}_k^l\| \geq T_l^H, & l = 1, \ldots, j-1, j+1, \ldots, m, \end{cases} \tag{2.80}$$

while $T_i^H$ denotes a prespecified threshold.

Similarly to the case of the sensor fault isolation scheme, in order to design the actuator fault isolation scheme it is necessary to assume that all sensors are fault free. Moreover, the term $\boldsymbol{h}(\boldsymbol{u}_k)$ in

$$\boldsymbol{x}_{k+1} = \boldsymbol{g}(\boldsymbol{x}_k) + \boldsymbol{h}(\boldsymbol{u}_k) + \boldsymbol{E}_k \boldsymbol{d}_k, \tag{2.81}$$

$$\boldsymbol{y}_{k+1} = \boldsymbol{C}_{k+1} \boldsymbol{x}_{k+1} \tag{2.82}$$

should have the following structure:

$$\boldsymbol{h}(\boldsymbol{u}_k) = \boldsymbol{B}(\boldsymbol{u}_k)\boldsymbol{u}_k, \tag{2.83}$$

where the $i$th column of $\boldsymbol{B}(\boldsymbol{u}_k)$ is a non-linear function of the form $\boldsymbol{b}_i(\boldsymbol{u}_k^i)$, and $\boldsymbol{u}_k^i \in \mathbb{R}^{r-1}$ is obtained from $\boldsymbol{u}_k$ by deleting its $i$th component, $u_{i,k}$.

In this case, for each of the observers the system can be characterised as follows:

$$\begin{aligned} \boldsymbol{x}_{k+1} &= \boldsymbol{g}(\boldsymbol{x}_k) + \boldsymbol{h}^i(\boldsymbol{u}_k^i + \boldsymbol{f}_k^i) + \boldsymbol{h}_i(u_{i,k} + f_{i,k}) + \boldsymbol{E}_k \boldsymbol{d}_k \\ &= \boldsymbol{g}(\boldsymbol{x}_k) + \boldsymbol{h}^i(\boldsymbol{u}_k^i + \boldsymbol{f}_k^i) + \boldsymbol{E}_k^i \boldsymbol{d}_k^i, \end{aligned} \tag{2.84}$$

$$\boldsymbol{y}_{k+1} = \boldsymbol{C}_{k+1} \boldsymbol{x}_{k+1}, \qquad i = 1, \ldots, r, \tag{2.85}$$

with

$$\boldsymbol{h}^i(\boldsymbol{u}_k^i + \boldsymbol{f}_k^i) = \boldsymbol{B}^i(\boldsymbol{u}_k)(\boldsymbol{u}_k^i + \boldsymbol{f}_k^i), \tag{2.86}$$

$$\boldsymbol{h}_i(u_{i,k} + f_{i,k}) = \boldsymbol{b}_i(\boldsymbol{u}_k^i)(u_{i,k} + f_{i,k}), \tag{2.87}$$

while $\boldsymbol{B}^i(\boldsymbol{u}_k)$ is obtained from $\boldsymbol{B}(\boldsymbol{u}_k)$ by deleting its $i$th column, and

$$\boldsymbol{E}_k^i = [\boldsymbol{E}_k \; \boldsymbol{b}_i(\boldsymbol{u}_k^i)], \quad \boldsymbol{d}_k^i = \begin{bmatrix} \boldsymbol{d}_k \\ u_{i,k} + f_{i,k} \end{bmatrix}. \tag{2.88}$$

Thus, the problem reduces to designing $r$ UIOs (Fig. 2.5). When all actuators but the $i$th one are fault-free, and all sensors are fault-free, then the residual $\boldsymbol{z}_k = \boldsymbol{y}_k - \hat{\boldsymbol{y}}_k$ will satisfy the following isolation logic:

$$\begin{cases} \|\boldsymbol{z}_k^i\| < T_i^H \\ \|\boldsymbol{z}_k^l\| \geq T_l^H, & l = 1, \ldots, i-1, i+1, \ldots, r. \end{cases} \tag{2.89}$$

The idea of an unknown input has also been employed in parity relation-based robust fault diagnosis for linear systems [42]. A robust parity relation method

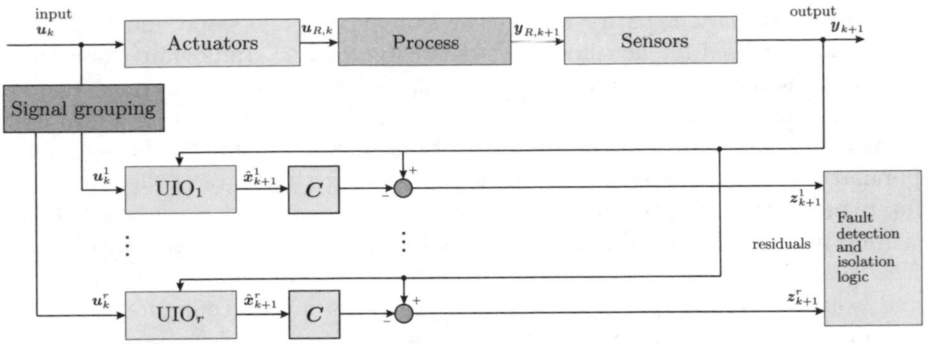


**Fig. 2.5.** Actuator fault detection and isolation scheme

for non-linear systems was proposed by Yu and Shields [192]. In particular, the bilinear class of non-linear systems was considered, which can be described as follows:

$$\boldsymbol{x}_{k+1} = \boldsymbol{A}\boldsymbol{x}_k + \sum_{i=1}^{r} \boldsymbol{A}^i u_{i,k}\boldsymbol{x}_k + \boldsymbol{B}\boldsymbol{u}_k + \boldsymbol{E}\boldsymbol{d}_k + \boldsymbol{L}_1\boldsymbol{f}_k, \tag{2.90}$$

$$\boldsymbol{y}_{k+1} = \boldsymbol{C}\boldsymbol{x}_{k+1} + \boldsymbol{D}\boldsymbol{u}_k + \boldsymbol{L}_2\boldsymbol{f}_{k+1}. \tag{2.91}$$

By including bilinear terms into the system matrix, a linear time-varying model of the following form is obtained:

$$\boldsymbol{x}_{k+1} = \boldsymbol{A}_k\boldsymbol{x}_k + \boldsymbol{B}\boldsymbol{u}_k + \boldsymbol{E}\boldsymbol{d}_k + \boldsymbol{L}_1\boldsymbol{f}_k, \tag{2.92}$$

$$\boldsymbol{y}_{k+1} = \boldsymbol{C}\boldsymbol{x}_{k+1} + \boldsymbol{D}\boldsymbol{u}_k + \boldsymbol{L}_2\boldsymbol{f}_{k+1}. \tag{2.93}$$

where

$$\boldsymbol{A}_k = \boldsymbol{A} + \sum_{i=1}^{r} \boldsymbol{A}^i u_{i,k}. \tag{2.94}$$

Yu and Shields proposed a recursive algorithm for calculating the matrices in the parity equation, which makes it possible to reduce the computational time significantly. They also showed how to perform an effective fault isolation with the singular value decomposition-based approach.

## 2.4 Concluding Remarks

The main objective of this chapter was to present the main principles of modern model-based fault diagnosis. Starting from elementary concepts and definitions, an outline of the most popular schemes for linear systems was presented (Section 2.1). In particular, three most popular approaches were considered, i.e., parameter estimation, parity relation and observers. The same line of presentation was realised for non-linear systems (Section 2.2). Important robustness issues of modern fault diagnosis were discussed in Section 2.3. In particular, it was shown how the robustness problem is tackled in the case of parameter estimation, observers, and the parity relation. It was also shown that the most popular way to settle such a challenging problem is to use the so-called unknown input model uncertainty. Indeed, this strategy has received a considerable attention in the fault diagnosis literature. This popularity can be explained by the fact that such an approach can be successfully used for robust fault detection and isolation. As was shown in Section 2.3, an appropriate configuration of the unknown input makes it possible to design a fault isolation scheme.

As can be observed in the literature, the most popular approach that uses unknown input model uncertainty is the so-called unknown input observer. Taking into account the presented state-of-the-art regarding observers and unknown input observers for non-linear systems, the number of real world applications (not only simple simulated systems) of non-linear observers should proliferate. Unfortunately, this is not the case. The main reason for such a situation is the relatively high design complexity of non-linear observers [2, 198]. This does not encourage engineers to apply those in industrial reality. Indeed, apart from the theoretically large potential of observer-based schemes, their computer implementations cause serious problems for engineers that, who usually are not fluent in the complex mathematical description involved in theoretical developments.

Taking into account the above problems, there are several tasks that has to be solved in order to make the unknown input observer for non-linear systems easier to implement in industrial reality:

Improve the convergence of linearisation-based techniques: it is an obvious fact that linearisation-based observers are almost as easy to implement as their linear counterparts. On the other hand, such approaches usually suffer from the lack of convergence. Thus, improving their convergence abilities is of paramount importance both from the theoretical and practical viewpoints.

Simplification of linearisation-free techniques: As was mentioned in Sections 2.2 and 2.3, there is a number of observers that can be used for particular classes of non-linear systems, e.g., Lipschitz systems, without the use of linearisation. Unfortunately, the design strategies of such observers are usually very complex, which limits their practical applications.

# 3. Soft Computing-Based FDI

Challenging design problems arise regularly in modern fault diagnosis systems. Unfortunately, the classic analytical techniques often cannot provide acceptable solutions to such difficult tasks. Indeed, as has already been mentioned, the design complexity of most observers for non-linear systems does not encourage engineers to apply those in practice. Another fact is that the application of observers is limited by the need for non-linear state-space models of the system being considered, which is usually a serious problem in complex industrial systems. This explains why most of the examples considered in the literature are devoted to simulated or laboratory systems, e.g., the known two- or three-tank system, the inverted pendulum, etc. [27, 96, 198]. Another serious difficulty is that there are examples for which fault directions are very similar to that of an unknown input. This may lead to a situation in which the effect of some faults is minimised and hence they may be impossible to detect. Other approaches that make use of the idea of an unknown input also inherit these drawbacks, e.g., the robust parity relation (see Section 2.3).

The above problems contribute to the rapid development of soft computing-based FDI [96, 132, 151, 148]. Generally, the most popular soft computing techniques that are used within the FDI framework can be divided into three groups:

- Neural networks;
- Fuzzy logic-based techniques;
- Evolutionary algorithms.

There are, of course, many combinations of such approaches, e.g., neuro-fuzzy systems [96, 99]. Another popular strategy boils down to integrating analytical and soft computing techniques, e.g., evolutionary algorithms and observers [183, 187] or neuro-fuzzy systems and observers [168].

The material of this book is limited to the so-called quantitative soft computing techniques, namely, neural networks and evolutionary algorithms [181]. Because of the fact that fuzzy logic-based approaches can be perceived as qualitative soft computing tools, they are beyond the scope of this book. For a comprehensive description regarding such techniques, the reader is referred to [96].

M. Witczak: Model. and Estim. Strat. for Fault Diagn. of Non-Linear Syst. LNCIS 354, pp. 31–46, 2007.
springerlink.com 

The chapter is organised as follows: Section 3.1.1 presents an elementary background regarding neural networks. In particular, starting from simple neural networks for static systems up to complex structures used for non-linear dynamic systems, the advantages and drawbacks of neural networks are discussed. Section 3.1.2 presents a bibliographical review regarding the application of neural networks to FDI.

A similar line of presentation is realised for evolutionary algorithms. In particular, Section 3.2 presents an elementary background regarding evolutionary algorithms while Section 3.2.2 presents a bibliographical review regarding their application to FDI.

## 3.1 Neural Networks

Generally, neural networks [73] can be perceived as a conveniently parameterised set of non-linear maps. In the last fifteen years, neural networks have been successfully used for solving complex problems in modelling and pattern recognition (see [73] and the references therein). In the case of pattern recognition, a finite set of input-output pairs is given, where the inputs represent the objects to be recognised while the outputs stand for the pattern classes to which they belong. Thus, the role of a neural network is to approximate the map between these two spaces. In the case of modelling, it is assumed that the input-output relation is formed by a non-linear system, and the role of a neural network is to approximate the behaviour of this system. In both cases, the application of neural networks is justified by the assumption that there exists a non-linear input-output map. The key theoretical result behind both applications is the fact that neural networks are universal approximators [73]. There are, of course, many different properties (see, e.g., [25]) which make neural networks attractive for practical applications.

One objective of the subsequent part of this section is to present the most frequently used structures of neural networks and to outline their advantages and drawbacks (Section 3.1.1). Another objective is to present a general view on the problem of using neural networks in fault diagnosis schemes. In particular, Section 3.1.2 presents a bibliographical review regarding the application of neural networks to FDI.

### 3.1.1 Basic Structures

This section reviews the well-known and frequently used Artificial Neural Networks (ANNs), which can be employed to the identification of static non-linear systems. In particular, the so-called feed-forward networks such as the Multi-Layer Perceptron (MLP) and Radial Basis Function (RBF) networks are considered.

#### Multi-layer perceptron

Artificial neural networks consist of a number of sub-units called neurons. The classic neuron structure (Fig. 3.1) can be described by

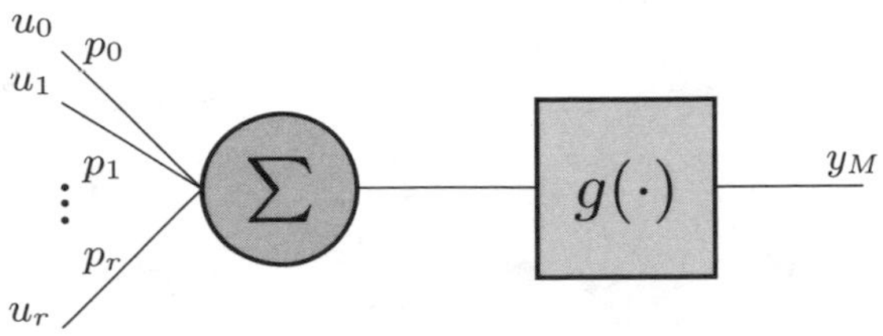


**Fig. 3.1.** Classic neuron structure

$$y_M = g\left(\sum_{i=0}^{r} p_i u_i\right) = g(\boldsymbol{p}^T \boldsymbol{u}), \text{ with } u_0 = 1, \tag{3.1}$$

where $g(\cdot)$ stands for the so-called activation function, and, as usual, $\boldsymbol{p}$ is the parameter (or weight) vector to be estimated. It is obvious that the behaviour of the neuron (3.1) depends on the activation function $g(\cdot)$. There are, of course, many different functions which can be employed to settle this problem. The simplest choice is to use a linear activation function resulting in the so-called *Adaline* neuron [76]. Nevertheless, real advantages of neural networks can be fully exploited when activation functions are non-linear. Typically, the activation function is chosen to be of the saturation type. The common choice is to use sigmoidal functions such as the logistic

$$g(x) = \text{logistic}(x) = \frac{1}{1 + \exp(-x)} \tag{3.2}$$

and the hyperbolic tangent

$$g(x) = \tanh(x) = \frac{1 - \exp(-2x)}{1 + \exp(-2x)} = 2\text{logistic}(2x) - 1 \tag{3.3}$$

functions. As can be seen from (3.3), the functions can be transformed into each other. Moreover, these two functions share an interesting property, namely, that their derivatives can be expressed as simple functions of the outputs. As long as a gradient-based algorithm is used to obtain parameter estimates, this property leads to a significant decrease in the computational burden, thus making the network synthesis process more effective.

The multi-layer perceptron is a network consisting of neurons divided into the so-called layers (Fig. 3.2). Such a network possesses an input layer, one or more hidden layers, and an output layer. The main tasks of the input layer are data preprocessing (e.g., scaling, filtering, etc.) and passing input signals into the hidden layer. Therefore, only the hidden and output layers constitute a "true" model. The connections between neurons are designed in such a way that each neuron of the former layer is connected with each element of the succeeding one. The non-linear neural network model (cf. Fig. 3.2) can symbolically be expressed as follows:

$$y_M = \boldsymbol{g}_3\left(\boldsymbol{g}_2\left(\boldsymbol{g}_1\left(\boldsymbol{u}\right)\right)\right), \tag{3.4}$$

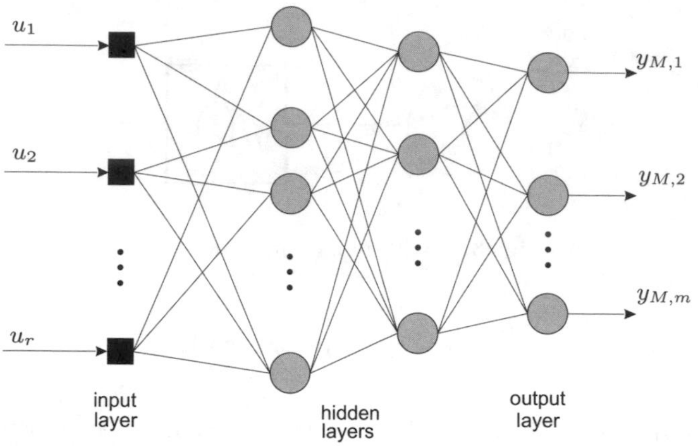


**Fig. 3.2.** Exemplary multi-layer perceptron with 3 layers

where $\boldsymbol{g}_1(\cdot)$, $\boldsymbol{g}_2(\cdot)$, and $\boldsymbol{g}_3(\cdot)$ stand for operators defining signal transformation through the 1st, 2nd and output layers, respectively.

One of the fundamental advantages of neural networks is their learning and adaptational abilities. An MLP network is a *universal approximator* [78]. This means that the MLP can approximate any smooth function with an arbitrary degree of accuracy as the number of hidden layer neurons increases. From the technical point of view, the training of neural networks is nothing else but parameter estimation. Indeed, once the structure of a network is known, the remaining task is to obtain the parameter vector $\boldsymbol{p}$. To tackle this problem, the celebrated backpropagation algorithm [43, 76] can be employed. Other possibilities involve the application of various stochastic [137, 170] or evolutionary algorithms [43]. These approaches should be adapted when the classic gradient-based algorithms fail to converge to satisfactory results. This is, however, a common situation, owing to the multimodal character of the optimisation index. Another problem may occur because of a large number of parameters to be estimated. This is especially true for neural networks with many hidden layers.

The main drawback of neural networks arises from model structure selection. There are, of course, many more or less sophisticated approaches to this problem, and they can be divided into three classes:

1. Bottom-up approaches: starting with a relatively simple structure, the number of hidden neurons is increased.
2. Top-down approaches: starting from a relatively complex structure, which seems to be sufficient to solve an identification problem, the "excessive" neurons are removed.
3. Discrete optimisation methods: with each network structure, an evaluation value is associated and then the network structure space is explored to find an appropriate configuration.

Unfortunately, the efficiency of those algorithms is usually very limited. As a result, neural networks with very poor generalisation abilities are obtained.

Another drawback of neural networks is the fact that models resulting from this approach are not in a "human readable" form. This means that the structure of such a network is not able to provide any practical knowledge about that of a real system. In fact, according to the literature, neural networks are called *black boxes*.

### Radial basis function networks

Radial basis function networks as rivals of arduously learning multi-layer perceptrons have received considerable research attention in the recent years [43, 123]. This kind of networks requires many nodes to achieve satisfactory results. The problem is somewhat similar to that of selecting an appropriate structure of a multi-layer perceptron. The RBF network (Fig. 3.3) consists of three layers, namely, the input layer, one hidden layer, and the output layer. The output $\phi_i$ of the $i$th neuron of the hidden layer is a non-linear function of the Euclidean distance from the input vector $\boldsymbol{u}$ to the vector of centres $\boldsymbol{c}_i$, and can be expressed as follows:

$$\phi_i = g\left(\|\boldsymbol{u} - \boldsymbol{c}_i\|_2\right), \; i = 1, \ldots, n_h, \tag{3.5}$$

where $\|\cdot\|_2$ stands for the Euclidean norm and $n_h$ is the number of neurons in the hidden layer. The $j$th network output is a weighted sum of the hidden neurons' output:

$$y_{M,j} = \sum_{i=1}^{n_h} p_{ji}\phi_i. \tag{3.6}$$

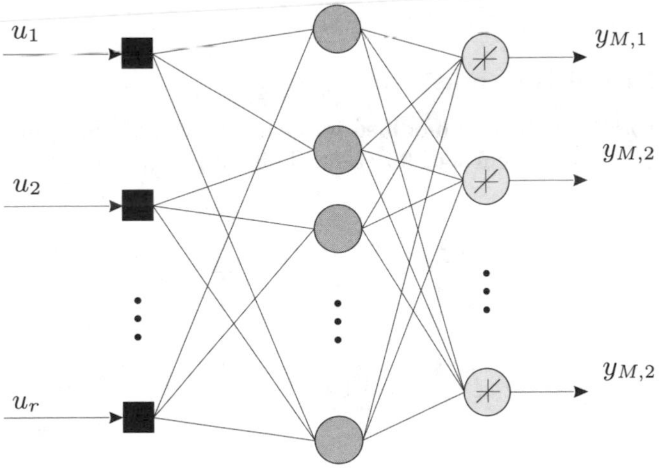


**Fig. 3.3.** Exemplary radial basis function networks

The activation function $g(x)$ is usually chosen to possess a maximum at $x = 0$. Typical choices for the activation function are the Gaussian function,

$$g(x) = \exp\left(-\frac{x^2}{\rho^2}\right), \tag{3.7}$$

and the inverse multi-quadratic function,

$$g(x) = \frac{1}{\sqrt{x^2 + \rho^2}}, \tag{3.8}$$

where $\rho$ signifies an additional free parameter.

The fundamental task in designing RBF networks is the selection of the number of hidden neurons and the activation function type. Then, function centres and their positions should be chosen. In this context, it should be pointed out that too small a number of centres may result in poor approximation properties.

On the other hand, the number of exact centres increases according to the dimension of the input space. For the sake of this, the application of RBF networks is rather restricted to low-dimensional problems.

A typical strategy for RBF network training consists in exploiting the linearity of the output layer parameters (weights) and geometric interpretability of the hidden layer parameters. The hidden layer parameters are determined first and, subsequently, the output layer parameters are obtained by means of some well-known approaches for linear parameter estimation, e.g., by the least-square algorithm. There are, of course, many more or less sophisticated approaches to selecting the centres and the widths of the basis function. The simplest one consists in randomly selecting these parameters; however, this is not a really practical approach. More efficient and, of course, more sophisticated approaches rely on the application of clustering, grid-based and subset selection techniques, as well as non-linear optimisation (see [123] for a survey).

## Dynamic Neural Networks

Neural networks can also be modified in such a way that they can be useful to identify a dynamic system [86, 96, 123]. This can be achieved by introducing

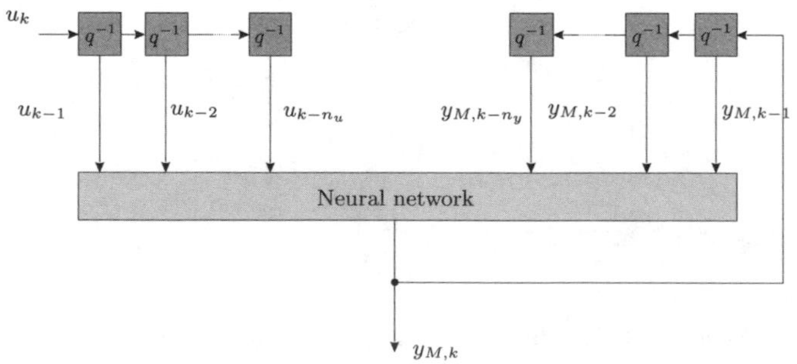


**Fig. 3.4.** Neural network with tapped delay lines

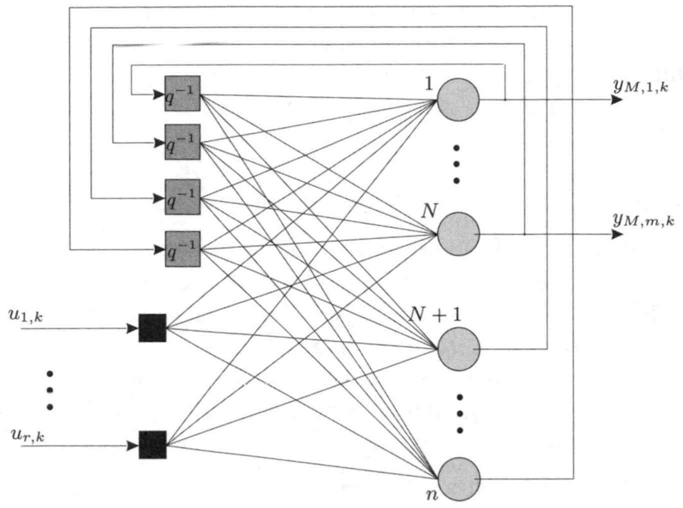


**Fig. 3.5.** Recurrent neural network developed by Williams and Zipser

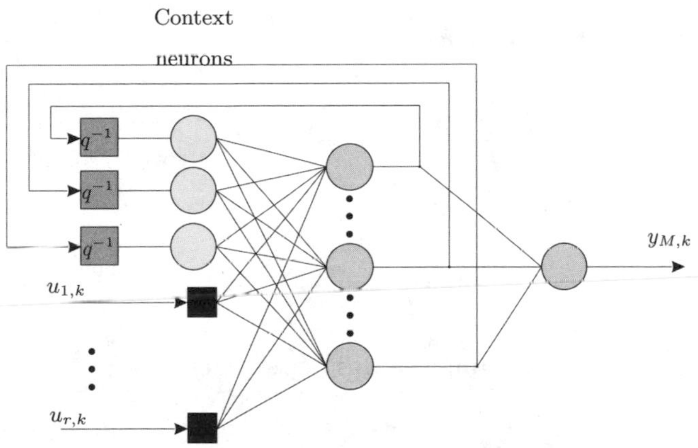


**Fig. 3.6.** Partially recurrent Elman neural network

tapped delay lines into the model (Fig. 3.4). The multi-layer perceptron or the radial basis function network can be employed as the main part of the overall model. A recurrent network developed by Williams and Zipser [177] consists of $n$ fully connected neurons, $r$ inputs and $m$ outputs (Fig. 3.5). As can be clearly seen, such a network has no feed-forward architecture. The main disadvantages of such networks are caused by the slow convergence of the existing training algorithms as well as stability problems of the resulting model.

Contrary to the above-mentioned fully recurrent structures, partially recurrent networks are based on feed-forward multi-layer perceptrons containing the so-called context layer, as in the case of the Elman neural network (Fig. 3.6) [123].

In such a network, feedback connections are realised from the hidden or output layers to the context neurons. The recurrency is more structured, which leads to faster training. As in the case of recurrent networks and most other approaches, the disadvantages of partially recurrent neural networks arise from model order selection as well as stability.

### Locally recurrent globally feed-forward networks

In the case of fully or partially recurrent neural networks, either all or selected connections are allowable. All neurons are similar to those of static networks, i.e., they are static (no feedback within a neuron). Those global connections cause various disadvantages, e.g., the lack of stability. An alternative solution is to introduce dynamic neurons into the feed-forward network. There are, of course, many different neuron models which can be employed for that purpose. The best known architectures are the following:

- Neurons with local activation feedback [48]:

$$y_{M,k} = g\left(y_{l,k}\right), \; y_{l,k} = \sum_{i=1}^{r} p_i u_{i,k} + \sum_{i=1}^{n_y} p_{r+i} y_{l,k-i}. \tag{3.9}$$

- Neurons with local synapse feedback [10]:

$$y_{M,k} = g\left(\sum_{i=1}^{r} G_i(q) u_{i,k}\right), \; G_i(q) = \frac{\sum_{j=0}^{n_u} b_j q^{-1}}{\sum_{j=0}^{n_y} a_j q^{-1}}. \tag{3.10}$$

- Neurons with output feedback [68] :

$$y_{M,k} = g\left(\sum_{i=1}^{r} p_i u_{i,k} + \sum_{i=1}^{n_y} p_{r+i} y_{M,k-i}\right). \tag{3.11}$$

- Neurons with an Infinite Impulse Response (IIR) filter [9, 137, 139]:

$$y_k = g\left(y_{l,k}\right), \; y_{l,k} = \sum_{i=0}^{n_u} b_i s_{k-1} + \sum_{i=1}^{n_y} a_i y_{l,k-1},$$
$$s_k = \sum_{i=1}^{r} p_i u_{i,k}. \tag{3.12}$$

The main advantage of locally recurrent globally feed-forward networks is that their stability can be proven relatively easily. As a matter of fact, the stability of the network depends only on the stability of the neurons. In most cases the stability conditions of a neuron boil down to checking the stability of a linear submodel. The feed-forward structure of such networks seems to make the training process easier. On the other hand, the introduction of dynamic neurons enlarges the parameter space significantly. This drawback together with the non-linear and multi-modal properties of an identification index implies that parameter estimation (or training) becomes relatively complex.

### 3.1.2 Neural Networks in FDI

At the beginning of the 1990s, neural networks were proposed for identification and control (see, e.g., [122]). The rapid development concerning applications of neural networks to control engineering resulted in a vast number of publications related to this subject. In 1992, Hunt *et al.* [83] confirmed the fast development of this research area by publishing a survey on neural networks in control engineering. In 1995, a similar work was published by Sjoberg *et al.* [160] in the context of system identification with neural networks. Nowadays, the vast number of applications has increased significantly. Fault diagnosis constitutes one of the thrusts of the research effort on neural networks for control [96].

The main objective of the subsequent part of this section is to present the development of this particular research area. Rather than providing an exhaustive survey on neural networks in fault diagnosis, it is aimed at providing a comprehensive account of the published work that exploits the special nature of neural networks. Indeed, it is impossible to count all publications on fault diagnosis in which neural networks are used as models of the systems being diagnosed. The strategy underlying such an approach boils down to generating the residual with the system and neural network output according to the simple residual generation scheme presented in Fig. 1.4. Examples of using such an approach with the classic multi-layer perceptron are leakages detection in an electro-hydraulic cylinder drive in a fluid power system [174], the diagnosis of non-catastrophic faults in a nuclear plant [175], and process valve actuator fault diagnosis [89]. Similar examples, but with dynamic neural networks, are the diagnosis of a chemical plant [58], the diagnosis of a valve actuator [96, 137], and the diagnosis of a steam evaporator [86]. Neural networks are also immensely popular in control and fault diagnosis schemes for induction motors [90, 64, 129, 148]. Finally, there are works that deal with fault diagnosis of transmission lines [131] and analog circuits [130].

There is a number of works concerning observer design with neural networks [3, 72]. Thus, when non-linear state-space models are available, then these approaches can be utilised for residual generation and fault diagnosis. Moreover, robustness with respect to model uncertainty can also be realised by using the concept of an unknown input. As has already been mentioned, when the direction of faults is similar to that of an unknown input, then the unknown input decoupling procedure may considerably impair fault sensitivity. If the above-mentioned approach fails, then describing model uncertainty in a different way seems to be a good remedy. One of the possible approaches is to use statistical techniques [7, 170] (for an example regarding different approaches, the reader is referred to [40]) to obtain parameter uncertainty of the model and, consequently, model output uncertainty. Such parameter uncertainty is defined as the parameter confidence region [7, 170] containing a set of admissible parameters that are consistent with the measured data. Thus it is evident that parameter uncertainty depends on measurement uncertainty, i.e., noise, disturbances, etc.

The knowledge about parameter uncertainty makes it possible to design the so-called adaptive threshold [57]. The adaptive threshold, contrary to the fixed

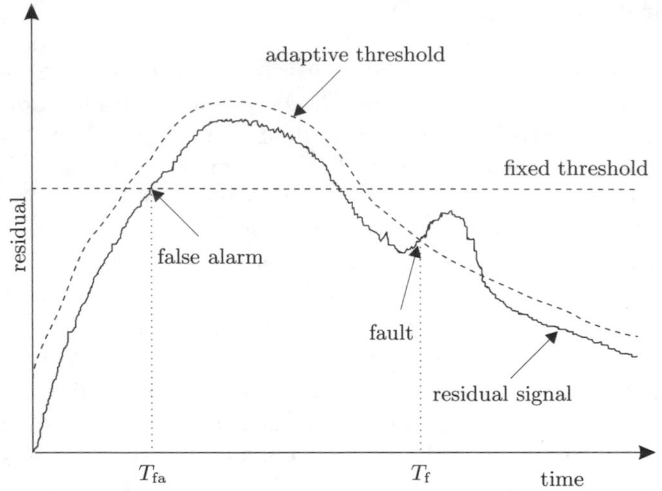


**Fig. 3.7.** Principle of an adaptive threshold

one (cf. Fig. 3.7), bounds the residual at a level that is dependent on model uncertainty, and hence it provides a more reliable fault detection. Contrary to the typical industrial applications of neural networks that are presented in the literature [27, 89, 96], Witczak *et al.* [185] defined the task of designing a neural network in such a way as to obtain a model with a possibly small uncertainty. Indeed, the approaches presented in the literature try to obtain a model that is best suited to a particular data set. This may result in a model with a relatively large uncertainty. A degraded performance of fault diagnosis constitutes a direct consequence of using such models. To tackle this challenging problem, the authors adapted and modified the GMDH (Group Method of Data Handling) approach [85, 96]. They proposed a complete design procedure concerning the application of GMDH neural networks to robust fault detection. Starting from a set of input-output measurements of the system, it is shown how to estimate the parameters and the corresponding uncertainty of a neuron using the so-called bounded-error approach [116, 170]. As a result, they obtained a tool that is able to generate an adaptive threshold. The methodology developed for parameter and uncertainty estimation of a neuron makes it possible to formulate an algorithm that allows obtaining a neural network with a relatively small modelling uncertainty. All the hard computations regarding the design of the GMDH neural network are performed off-line, and hence the problem regarding the time-consuming calculations is not of paramount importance. The approach can also be extended for dynamic systems by using the dynamic neuron structure [136]. The above-mentioned technique will be clearly detailed in Section 7.2.

It is well known that the reliability of such fault diagnosis schemes is strongly dependent on model uncertainty, i.e., the mismatch between a neural network and the system being considered. Thus, it is natural to minimise model uncertainty as far as possible. This can be realised with the application of Optimum

Experimental Design (OED) theory [7, 167, 170]. Some authors have conducted active investigations in this important research area. White [176], MacKay [108], and Cohn [35] showed the attractiveness of the application of OED to neural networks. Fukumizu [59, 61] developed the so-called statistical active learning technique, which is based on the general theory of OED. Recently, Witczak and Prętki [188] developed a D-optimum experimental design strategy that can be used for training single-output neural networks. They also showed how to use the obtained network for robust fault detection with an adaptive threshold. In [182], the author showed how to extend this technique to multi-input multi-output neural networks. He also proposed a sequential experimental design algorithm that allows obtaining a one-step-ahead D-optimum input. This algorithm can be perceived as a hybrid one since it can be used for both training and data development. Section 7.1 presents all the details regarding the above-described design methodology.

Finally, there is also a large number of approaches that use neural networks as pattern classifiers [96] to tackle the FDI problem. Instead of using neural networks as the models of the systems being diagnosed, the networks are trained to recognise different modes of the system, i.e., both faulty and non-faulty ones. Examples of using such an approach are FDI in hydraulic fluid power systems [105, 106], FDI in machine dynamics and vibration problems [190], sensor fault diagnosis [194], fault diagnosis of chemical processes [197], and fault diagnosis of a two-tank system [96].

## 3.2 Evolutionary Algorithms

Evolutionary Algorithms (EAs) are a broad class of stochastic optimisation algorithms inspired by some biological processes, which allow populations of organisms to adapt to their surrounding environment. Such algorithms have been influenced by Darwin's theory of natural selection, or the survival of the fittest (published in 1859). The idea behind it is that only certain organisms can survive, i.e., only those which can adapt to the environment and win the competition for food and shelter. Almost at the same time that Darwin's theory was presented (1865), Mendel published a short monograph about experiments with plant hybridisation. He observed how traits of different parents are combined into offspring by sexual reproduction. Darwinian evolutionary theory and Mendel's investigations of heredity in plants became the foundations of evolutionary search methods and led to the creation of the neo-Darwinian paradigm [53].

One objective of the subsequent part of this section is to present the main principles of evolutionary algorithms and to describe their typical forms (Section 3.2.1). Another objective is to present a general view on the problem of using evolutionary algorithms in fault diagnosis schemes. In particular, Section 3.2.2 presents a bibliographical review regarding the application of evolutionary algorithms to FDI.

### 3.2.1 Introduction to Evolutionary Computation

In order to give a general outline of an evolutionary algorithm, let us introduce a few different concepts and notations [115].

An evolutionary algorithm is based on a collective learning process within *a population* of $n_{\text{pop}}$ individuals, each of which represents *a genotype* (an underlying genetic code), a search point in the so-called *genotype space*. The environment delivers quantitative information (the fitness value) regarding an individual based on its *phenotype* (the manner of response contained in the behaviour, physiology and morphology of an organism). Thus, each individual has its own phenotype and genotype.

The general principle of an evolutionary algorithm can be described as follows: At the beginning, a population is randomly initialised and evaluated, i.e., based on a phenotype, the fitness of each individual is calculated. Next, randomised processes of *reproduction*, *recombination*, *mutation* and *succession* are iteratively repeated until a given termination condition is reached. *Reproduction*, called also *preselection*, is a randomised process (deterministic in some algorithms) of parent selection from the entire population, i.e., a temporary population of parent individuals is formed. The *recombination* mechanism (omitted in some algorithms) allows mixing parental information while passing it to the descendants. *Mutation* introduces an innovation into the current descendants. Finally, *succession*, called also *post selection*, is applied to choose a new generation of individuals from parents and descendants. All the above operations are repeated until the termination condition is reached.

This is, of course, a general principle, and it can be more or less modified for various types of evolutionary algorithms.

The duality of genotype and phenotype suggests two main approaches to simulated evolution [115]. In genotypic simulations, the attention focuses on genetic structures. This means that the entire searching process is performed in the genotype space. However, in order to calculate the individual's fitness, its chromosome must be decoded to its phenotype. Nowadays, two kinds of such algorithms can be distinguished, i.e.,

- Genetic Algorithms (GAs) [77];
- Genetic Programming (GP) [103].

In phenotypic simulations, the attention focuses on the behaviour of candidate solutions in a population. All operations, i.e., selection, reproduction, and mutation are performed in the phenotype space. Nowadays, three main kinds of such algorithms can be distinguished, i.e.,

- Evolutionary programming [54];
- Evolutionary strategies [115];
- Evolutionary Search with Soft Selection (ESSS) [63, 126].

Special attention in this section is focused on the genetic programming strategy as this is one of the main tools employed in Part III. First, let us introduce some elementary background on genetic algorithms.

Genetic algorithms are computation models that approach the natural evolution perhaps most closely. Many works confirm their effectiveness and recommend their application to various optimisation problems.

A genetic algorithm processes a population of individuals whose DNA is represented by fixed-length binary strings. Inside a computer programme, an individual's fitness is calculated directly from the DNA, and so only the DNA has to be represented. The population of such individuals is evolved through successive generations; individuals in each new generation are bred from the fittest individuals from the previous generation.

The breeding of a new parent is inspired by natural processes, i.e., either asexual or sexual reproduction can be employed. In asexual reproduction, the parent individual is simply copied (possibly with some random changes within a genotype). This process is called *mutation* (Fig. 3.8). In sexual reproduction, couples of parents are randomly chosen and new individuals are created by alternately copying sequences from each parent. This process is known as *crossover* (Fig. 3.9).

The main difference between GAs and genetic programming is that in GP the evolving individuals are parse trees rather than fixed-length binary strings (cf. Fig. 3.10). Genetic programming applies the approach of GAs to a population of programs which can be described by such trees. Such an approach has demonstrated its potential by evolving simple programs for medical signal

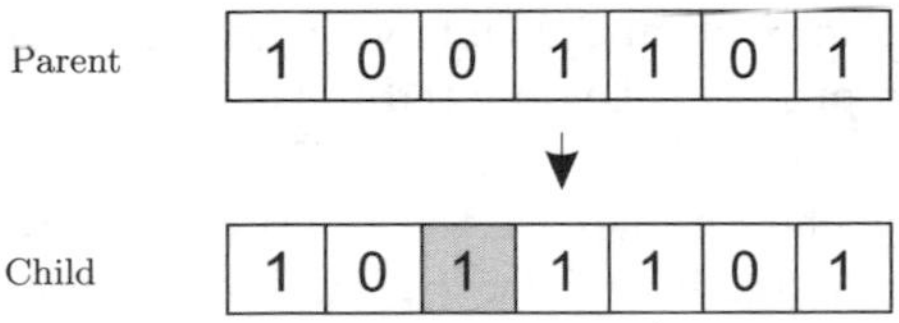


**Fig. 3.8.** Exemplary mutation operation

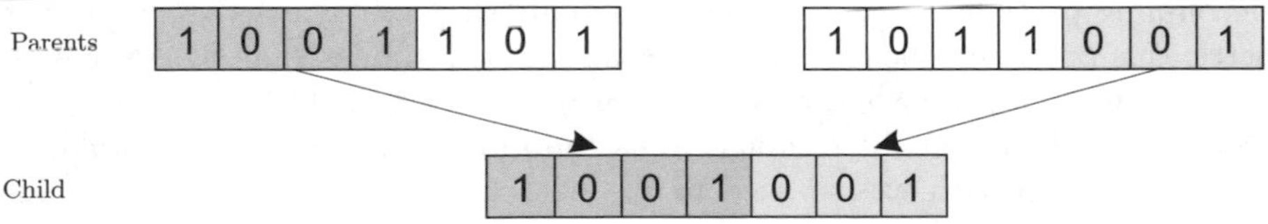


**Fig. 3.9.** Exemplary crossover operation

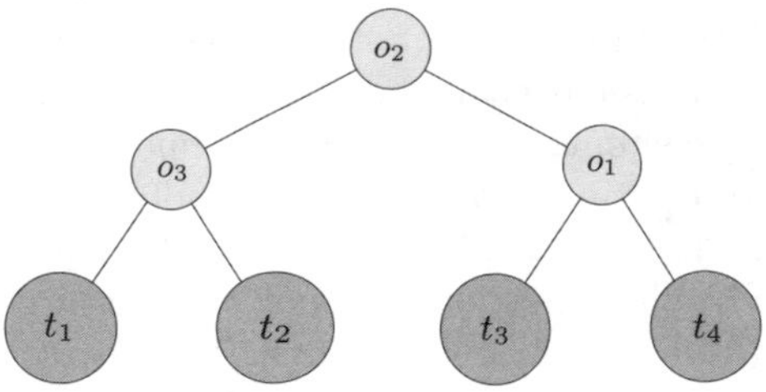


**Fig. 3.10.** Exemplary GP tree

filters, or by performing optical character recognition, target identification, system identification, fault diagnosis, etc. [47, 69, 103, 183, 187].

### 3.2.2 Evolutionary Algorithms in FDI Schemes

Although the origins of evolutionary algorithms can be traced back to the late 1950s (see [11] for a comprehensive introduction and survey on EAs), the first works on evolutionary algorithms in control engineering were published at the beginning of the 1990s. In 2002, Fleming and Purshouse [51] tackled a challenging task of preparing a comprehensive survey on the application of evolutionary algorithms to control engineering. As is indicated in [51], there are relatively scarce publications on applications of evolutionary algorithms to the design of FDI systems.

This section, rather than providing an exhaustive survey on evolutionary algorithms in fault diagnosis, is aimed at providing a comprehensive account of the published works that exploit the special nature of EAs [181]. This means that the works dealing with EAs applied as alternative optimisers, e.g., for training neural and/or fuzzy systems are not included here. In other words, the main objective is to extend the material of [51] by introducing the latest advances in fault diagnosis with evolutionary algorithms.

As was mentioned in Section 2.3, many approaches have been proposed to tackle the robustness problem. Undoubtedly, the most common approach is to use robust observers, such as the UIO [27, 96, 179], which can tolerate a degree of model uncertainty and hence increase the reliability of fault diagnosis. In such an approach, the model-reality mismatch can be represented by the so-called unknown input. Hence the state estimate and, consequently, the output estimate are obtained taking into account model uncertainty. Unfortunately, when the direction of faults is similar to that of an unknown input, then the unknown input decoupling procedure may considerably impair the fault sensitivity. In order to settle this problem, Chen *et al.* [28] (see also [27]) formulated an observer-based FDI as a multiobjective optimisation problem, in which the task was to maximise the effect of faults on the residual, whilst minimising the effect of an unknown input. The approach was applied to the detection of sensor faults in a flight control system. A similar approach was proposed by Kowalczuk *et al.* [102], where the observer design is founded on a Pareto-based approach, in which the ranking of an individual solution is based on the number of solutions by which it is dominated. These two solutions can be applied to linear systems only.

In spite of the fact that a large amount of knowledge on designing observers for non-linear systems has been accumulated through the literature since the beginning of the 1970s, a customary approach is to linearise the non-linear model around the current state estimate, and then to apply techniques for linear systems, as is the case for the extended Kalman filter [96]. Unfortunately, this strategy works well only when linearisation does not cause a large mismatch between the linear model and the non-linear behaviour. To improve the effectiveness of state estimation, it is necessary to restrict the class of non-linear systems while designing observers. Unfortunately, the analytical design procedures resulting from such an approach are usually very complex, even for simple laboratory systems [198]. To overcome this problem, Porter and Passino proposed the so-called genetic adaptive observer [144]. They showed how to construct such an observer where a genetic algorithm evolves the gain matrix of the observer in real time so that the output error is minimised. Apart from the relatively simple design procedure, the authors did not provide the convergence conditions of the observer. They did not consider robustness issues with respect to model uncertainty. A solution that does not possess such drawbacks was proposed by Witczak *et al.* [187]. In particular, the authors showed the convergence condition of the observer and proposed a technique for increasing its convergence rate with genetic programming. This approach will be detailed in Section 6.2.

It should be strongly underlined that the application of observers is limited by the need for non-linear state-space models of the system being considered, which is usually a serious problem in complex industrial systems. This explains why most of the examples considered in the literature are devoted to simulated or laboratory systems, e.g., the celebrated three- (two- or even four-) tank system, an inverted pendulum, a travelling crane, etc. To tackle this problem, a genetic programming-based approach for designing state-space models from input-output data was developed in [114, 179, 187]. This approach will be detailed in Section 6.1. A further development of this technique related to input-output models can be found in [114].

Evolutionary algorithms have also been applied to FDI methods that are not based on the concept of residuals. Marcu [110] formulated FDI design as a feature selection and classifier design problem. EA has also been applied to the generalised task of determining the fault from a collection symptoms [117]. The method relied upon the availability of a priori probabilities that a particular fault caused a particular symptom. In [30], the authors employed a genetic algorithm-based evolutionary strategy for fault diagnosis-related classification problems, which includes two aspects: evolutionary selection of training samples and input features, and evolutionary construction of the neural network classifier. Finally, Sun *et al.* [162] used the bootstrap technique to preprocess the operational data acquired from a running diesel engine, and a genetic programming approach to find the best compound feature that can discriminate among four kinds of commonly operating modes of the engine.

## 3.3 Concluding Remarks

The main objective of this chapter was to present quantitative soft computing approaches to fault diagnosis. In particular, elementary neural network structures for both static and dynamic non-linear systems were presented (Section 3.1.1). Moreover, their application to FDI was discussed based on the bibliographical review presented in Section 3.1.2. A similar line of presentation was realised for evolutionary algorithms in Sections 3.2 and 3.2.2.

There is no doubt that soft computing techniques can be very attractive tools for solving various problems related to modern FDI. On the other hand, their application usually involves a high computational burden. This means that soft computing techniques should be used only in justified situations. Such a situation usually takes place when the classic analytical techniques cannot be employed or they do not provide satisfactory results.

Apart from the variety of different FDI applications of soft computing methods, two general approaches deserve special attention, namely,

Integration of analytical and soft computing FDI techniques: there is a number of design strategies for analytical techniques, e.g., [27, 96, 187] whose performance can be significantly improved by the use of soft computing techniques.

Robust soft computing-based FDI techniques: this mainly concerns artificial neural networks. Contrary to the industrial applications of neural networks that are presented in most of the published books and papers, the task of designing a neural network is defined in such a way as to obtain a model with a possibly small uncertainty. Indeed, the approaches presented in the literature try to obtain a model that is best suited to a particular data set. This may result in a model with a relatively large uncertainty. The degraded performance of fault diagnosis constitutes a direct consequence of using such models. Another advantage of such approaches is the fact that they provide knowledge regarding model uncertainty that can be used to obtain the so-called adaptive threshold. Such a threshold enhances the performance of FDI and makes it robust to model uncertainty.

# Part II

# State and Parameter Estimation Strategies

# 4. State Estimation Techniques for FDI

The main objective of this chapter is to present three different observer structures that can be used for non-linear discrete-time systems with unknown inputs. As was pointed out in Chapter 2, there are two general tasks that have to be fulfilled while developing novel observer structures. The first one concerns the development of linearisation-based techniques and its main objective is to improve the convergence of such schemes. In order to solve this challenging problem, Section 4.1 introduces the concept of the so-called Extended Unknown Input Observer (EUIO) [96, 187]. This section presents a comprehensive convergence analysis of the EUIO with the Lyapunov method. Based on the achieved results, a complete design procedure is proposed and carefully described. Section 4.2 portrays further development of the EUIO. In particular, it exposes less restrictive convergence conditions than those proposed in Section 4.1. The achieved results are then used for the development of a new EUIO design procedure.

The second task underlined in Chapter 2 is related to the simplification of linearisation-free design procedures. Following this general requirement, Section 4.3 presents three alternative design procedures for observers and unknown input observers for Lipschitz non-linear discrete-time systems. In particular, the main objective is to present effective and simple to implement design procedures.

## 4.1 Extended Unknown Input Observer

As was mentioned in Section 2.3, the UIO can also be employed for linear stochastic systems. Although the primary purpose of the subsequent part of this section is to present an extended unknown input observer [183, 187], the information exhibited below is necessary to explain clearly the results of [187]. Following a common nomenclature, such an UIO will be called an Unknown Input Filter (UIF).

Let us consider the following linear discrete-time system:

$$\boldsymbol{x}_{k+1} = \boldsymbol{A}_k\boldsymbol{x}_k + \boldsymbol{B}_k\boldsymbol{u}_k + \boldsymbol{E}_k\boldsymbol{d}_k + \boldsymbol{L}_{1,k}\boldsymbol{f}_k + \boldsymbol{w}_k, \tag{4.1}$$

$$\boldsymbol{y}_k = \boldsymbol{C}_k\boldsymbol{x}_k + \boldsymbol{L}_{2,k+1}\boldsymbol{f}_{k+1} + \boldsymbol{v}_k. \tag{4.2}$$

M. Witczak: Model. and Estim. Strat. for Fault Diagn. of Non-Linear Syst. LNCIS 354, pp. 49–83, 2007.
springerlink.com

In this case, $\boldsymbol{v}_k$ and $\boldsymbol{w}_k$ are independent zero-mean white noise sequences. The matrices $\boldsymbol{A}_k$, $\boldsymbol{B}_k$, $\boldsymbol{C}_k$, $\boldsymbol{E}_k$ are assumed to be known and have appropriate dimensions. To overcome the state estimation problem of (4.1)–(4.2), an UIF with the following structure can be employed:

$$\boldsymbol{s}_{k+1} = \boldsymbol{F}_{k+1}\boldsymbol{s}_k + \boldsymbol{T}_{k+1}\boldsymbol{B}_k\boldsymbol{u}_k + \boldsymbol{K}_{k+1}\boldsymbol{y}_k, \tag{4.3}$$

$$\hat{\boldsymbol{x}}_{k+1} = \boldsymbol{s}_{k+1} + \boldsymbol{H}_{k+1}\boldsymbol{y}_{k+1}, \tag{4.4}$$

where

$$\boldsymbol{K}_{k+1} = \boldsymbol{K}_{1,k+1} + \boldsymbol{K}_{2,k+1}, \tag{4.5}$$

$$\boldsymbol{E}_k = \boldsymbol{H}_{k+1}\boldsymbol{C}_{k+1}\boldsymbol{E}_k, \tag{4.6}$$

$$\boldsymbol{T}_{k+1} = \boldsymbol{I} - \boldsymbol{H}_{k+1}\boldsymbol{C}_{k+1}, \tag{4.7}$$

$$\boldsymbol{F}_{k+1} = \boldsymbol{T}_{k+1}\boldsymbol{A}_k - \boldsymbol{K}_{1,k+1}\boldsymbol{C}_k, \tag{4.8}$$

$$\boldsymbol{K}_{2,k+1} = \boldsymbol{F}_{k+1}\boldsymbol{H}_k. \tag{4.9}$$

The above matrices are designed in such a way as to ensure unknown input decoupling as well as the minimisation of the state estimation error:

$$\boldsymbol{e}_{k+1} = \boldsymbol{x}_{k+1} - \hat{\boldsymbol{x}}_{k+1}. \tag{4.10}$$

As was mentioned in Section 2.3, the necessary condition for the existence of a solution to (4.6) is $\text{rank}(\boldsymbol{C}_{k+1}\boldsymbol{E}_k) = \text{rank}(\boldsymbol{E}_k) = q$ [27, p. 72, Lemma 3.1], and a special solution is

$$\boldsymbol{H}^*_{k+1} = \boldsymbol{E}_k\left[(\boldsymbol{C}_{k+1}\boldsymbol{E}_k)^T\boldsymbol{C}_{k+1}\boldsymbol{E}_k\right]^{-1}(\boldsymbol{C}_{k+1}\boldsymbol{E}_k)^T. \tag{4.11}$$

If the conditions (4.5)–(4.8) are fulfilled, then the fault-free, i.e., $\boldsymbol{f}_k = \boldsymbol{0}$ state estimation error is given by

$$\boldsymbol{e}_{k+1} = \boldsymbol{F}_{k+1}\boldsymbol{e}_k - \boldsymbol{K}_{1,k+1}\boldsymbol{v}_k - \boldsymbol{H}_{k+1}\boldsymbol{v}_{k+1} + \boldsymbol{T}_{k+1}\boldsymbol{w}_k. \tag{4.12}$$

In order to obtain the gain matrix $\boldsymbol{K}_{1,k+1}$, let us first define the state estimation covariance matrix:

$$\boldsymbol{P}_k = \mathcal{E}\left\{[\boldsymbol{x}_k - \hat{\boldsymbol{x}}_k][\boldsymbol{x}_k - \hat{\boldsymbol{x}}_k]^T\right\}. \tag{4.13}$$

Using (4.12), the update of (4.13) can be defined as

$$\begin{aligned}\boldsymbol{P}_{k+1} =& \boldsymbol{A}_{1,k+1}\boldsymbol{P}_k\boldsymbol{A}^T_{1,k+1} + \boldsymbol{T}_{k+1}\boldsymbol{Q}_k\boldsymbol{T}^T_{k+1} + \boldsymbol{H}_{k+1}\boldsymbol{R}_{k+1}\boldsymbol{H}^T_{k+1}\\ &- \boldsymbol{K}_{1,k+1}\boldsymbol{C}_k\boldsymbol{P}_k\boldsymbol{A}^T_{1,k+1} - \boldsymbol{A}_{1,k+1}\boldsymbol{P}_k\boldsymbol{C}^T_k\boldsymbol{K}^T_{1,k+1}\\ &+ \boldsymbol{K}_{1,k+1}\left[\boldsymbol{C}_k\boldsymbol{P}_k\boldsymbol{C}^T_k + \boldsymbol{R}_k\right]\boldsymbol{K}^T_{1,k+1},\end{aligned} \tag{4.14}$$

where

$$\boldsymbol{A}_{1,k+1} = \boldsymbol{T}_{k+1}\boldsymbol{A}_k. \tag{4.15}$$

To give the state estimation error $\boldsymbol{e}_{k+1}$ the minimum variance, it can be shown that the gain matrix $\boldsymbol{K}_{1,k+1}$ should be determined by

$$\boldsymbol{K}_{1,k+1} = \boldsymbol{A}_{1,k+1}\boldsymbol{P}_k\boldsymbol{C}_k^T\left[\boldsymbol{C}_k\boldsymbol{P}_k\boldsymbol{C}_k^T + \boldsymbol{R}_k\right]^{-1}. \tag{4.16}$$

In this case, the corresponding covariance matrix is given by

$$\boldsymbol{P}_{k+1} = \boldsymbol{A}_{1,k+1}\boldsymbol{P}'_{k+1}\boldsymbol{A}_{1,k+1}^T + \boldsymbol{T}_{k+1}\boldsymbol{Q}_k\boldsymbol{T}_{k+1}^T + \boldsymbol{H}_{k+1}\boldsymbol{R}_{k+1}\boldsymbol{H}_{k+1}^T, \tag{4.17}$$

$$\boldsymbol{P}'_{k+1} = \boldsymbol{P}_k - \boldsymbol{K}_{1,k+1}\boldsymbol{C}_k\boldsymbol{P}_k\boldsymbol{A}_{1,k+1}^T. \tag{4.18}$$

The above derivation is very similar to that which has to be performed for the classic Kalman filter [4]. Indeed, the UIF can be transformed to the KF-like form as follows:

$$\begin{aligned}\hat{\boldsymbol{x}}_{k+1} = {} & \boldsymbol{A}_k\hat{\boldsymbol{x}}_k + \boldsymbol{B}_k\boldsymbol{u}_k - \boldsymbol{H}_{k+1}\boldsymbol{C}_{k+1}[\boldsymbol{A}_k\hat{\boldsymbol{x}}_k + \boldsymbol{B}_k\boldsymbol{u}_k] \\ & - \boldsymbol{K}_{1,k+1}\boldsymbol{C}_k\hat{\boldsymbol{x}}_k - \boldsymbol{F}_{k+1}\boldsymbol{H}_k\boldsymbol{y}_k \\ & + [\boldsymbol{K}_{1,k+1} + \boldsymbol{F}_{k+1}\boldsymbol{H}_k]\boldsymbol{y}_k + \boldsymbol{H}_{k+1}\boldsymbol{y}_{k+1},\end{aligned} \tag{4.19}$$

or, equivalently,

$$\hat{\boldsymbol{x}}_{k+1} = \hat{\boldsymbol{x}}_{k+1/k} + \boldsymbol{H}_{k+1}\boldsymbol{\varepsilon}_{k+1/k} + \boldsymbol{K}_{1,k+1}\boldsymbol{\varepsilon}_k, \tag{4.20}$$

with

$$\hat{\boldsymbol{x}}_{k+1/k} = \boldsymbol{A}_k\hat{\boldsymbol{x}}_k + \boldsymbol{B}_k\boldsymbol{u}_k, \tag{4.21}$$

$$\boldsymbol{\varepsilon}_{k+1/k} = \boldsymbol{y}_{k+1} - \hat{\boldsymbol{y}}_{k+1/k} = \boldsymbol{y}_{k+1} - \boldsymbol{C}_{k+1}\hat{\boldsymbol{x}}_{k+1/k}, \tag{4.22}$$

$$\boldsymbol{\varepsilon}_k = \boldsymbol{y}_k - \hat{\boldsymbol{y}}_k. \tag{4.23}$$

The above transformation can be performed by substituting (4.4) into (4.3) and then using (4.7) and (4.8). As can be seen, the structure of the observer (4.20) is very similar to that of the Kalman filter. The only difference is the term $\boldsymbol{H}_{k+1}\boldsymbol{\varepsilon}_{k+1/k}$, which vanishes when no unknown input is considered.

### 4.1.1 Convergence Analysis and Design Principles

As has already been mentioned, the application of the EKF to the state estimation of non-linear deterministic systems has received considerable attention during the last two decades (see [19] and the references therein). This is mainly because the EKF can directly be applied to a large class of non-linear systems, and its implementation procedure is almost as simple as that for linear systems. Moreover, in the case of deterministic systems, the instrumental matrices $\boldsymbol{R}_k$ and $\boldsymbol{Q}_k$ can be set almost arbitrarily. This opportunity makes it possible to use them to improve the convergence of the observer, which is the main drawback of linearisation-based approaches. This section presents an extended unknown input observer for a class of non-linear systems which can be described by the following equations [187]:

$$\boldsymbol{x}_{k+1} = \boldsymbol{g}(\boldsymbol{x}_k) + \boldsymbol{h}(\boldsymbol{u}_k) + \boldsymbol{E}_k\boldsymbol{d}_k + \boldsymbol{L}_{1,k}\boldsymbol{f}_k, \tag{4.24}$$

$$\boldsymbol{y}_{k+1} = \boldsymbol{C}_{k+1}\boldsymbol{x}_{k+1} + \boldsymbol{L}_{2,k+1}\boldsymbol{f}_{k+1}, \tag{4.25}$$

where $\boldsymbol{g}(\boldsymbol{x}_k)$ is assumed to be continuously differentiable with respect to $\boldsymbol{x}_k$. Similarly to the EKF, the observer (4.20) can be extended to the class of non-linear systems (4.24)–(4.25). The algorithm presented below though can also, with minor modifications, be applied to a more general structure. Such a restriction is caused by the need for employing it for FDI purposes. This leads to the following structure of the EUIO:

$$\hat{\boldsymbol{x}}_{k+1/k} = \boldsymbol{g}(\hat{\boldsymbol{x}}_k) + \boldsymbol{h}(\boldsymbol{u}_k), \tag{4.26}$$

$$\hat{\boldsymbol{x}}_{k+1} = \hat{\boldsymbol{x}}_{k+1/k} + \boldsymbol{H}_{k+1}\boldsymbol{\varepsilon}_{k+1/k} + \boldsymbol{K}_{1,k+1}\boldsymbol{\varepsilon}_k. \tag{4.27}$$

It should also be pointed out that the matrix $\boldsymbol{A}_k$ used in (4.15) is now defined by

$$\boldsymbol{A}_k = \left.\frac{\partial \boldsymbol{g}(\boldsymbol{x}_k)}{\partial \boldsymbol{x}_k}\right|_{\boldsymbol{x}_k=\hat{\boldsymbol{x}}_k}. \tag{4.28}$$

### 4.1.2 Convergence of the EUIO

In this section, the Lyapunov approach is employed for convergence analysis of the EUIO. The approach presented here is similar to that described in [19], which was used in the case of the EKF-based deterministic observer. The main objective of this section is to show that the convergence of the EUIO strongly depends on an appropriate choice of the instrumental matrices $\boldsymbol{R}_k$ and $\boldsymbol{Q}_k$. Subsequently, the fault-free mode is assumed, i.e., $\boldsymbol{f}_k = \boldsymbol{0}$.

For notational convenience, let us define the a priori state estimation error:

$$\boldsymbol{e}_{k+1/k} = \boldsymbol{x}_{k+1} - \hat{\boldsymbol{x}}_{k+1/k}. \tag{4.29}$$

Substituting (4.24)–(4.25) and (4.26)–(4.27) into (4.10), one can obtain the following form of the state estimation error:

$$\boldsymbol{e}_{k+1} = \boldsymbol{e}_{k+1/k} - \boldsymbol{H}_{k+1}\boldsymbol{\varepsilon}_{k+1/k} - \boldsymbol{K}_{1,k+1}\boldsymbol{\varepsilon}_k. \tag{4.30}$$

As usual, to perform further derivations, it is necessary to linearise the model around the current state estimate $\hat{\boldsymbol{x}}_k$. This leads to the classic approximation:

$$\boldsymbol{e}_{k+1/k} \approx \boldsymbol{A}_k\boldsymbol{e}_k + \boldsymbol{E}_k\boldsymbol{d}_k. \tag{4.31}$$

In order to avoid the above approximation, the diagonal matrix $\boldsymbol{\alpha}_k = \mathrm{diag}(\alpha_{1,k}, \ldots, \alpha_{n,k})$ is introduced, which makes it possible to establish the following exact equality:

$$\boldsymbol{e}_{k+1/k} = \boldsymbol{\alpha}_k\boldsymbol{A}_k\boldsymbol{e}_k + \boldsymbol{E}_k\boldsymbol{d}_k, \tag{4.32}$$

and hence (4.30) can be expressed as

$$\begin{aligned}\boldsymbol{e}_{k+1} &= \boldsymbol{e}_{k+1/k} - \boldsymbol{H}_{k+1}\boldsymbol{C}_{k+1}\boldsymbol{e}_{k+1/k} - \boldsymbol{K}_{1,k+1}\boldsymbol{C}_k\boldsymbol{e}_k \\ &= [\boldsymbol{I} - \boldsymbol{H}_{k+1}\boldsymbol{C}_{k+1}][\boldsymbol{\alpha}_k\boldsymbol{A}_k\boldsymbol{e}_k + \boldsymbol{E}_k\boldsymbol{d}_k] - \boldsymbol{K}_{1,k+1}\boldsymbol{C}_k\boldsymbol{e}_k \\ &= [\boldsymbol{T}_{k+1}\boldsymbol{\alpha}_k\boldsymbol{A}_k - \boldsymbol{K}_{1,k+1}\boldsymbol{C}_k]\boldsymbol{e}_k. \end{aligned} \tag{4.33}$$

As can be observed in (4.33), the convergence of the EUIO depends strongly on $\boldsymbol{\alpha}_k$. Thus, the purpose of further deliberations is to determine conditions relating the convergence of the EUIO with $\boldsymbol{\alpha}_k$.

Let us start with the following assumptions:

**Assumption 4.1.** Following [27] and, then [19], it is assumed that the system is locally uniformly rank observable. This guarantees (see [19] and the references therein) that the matrix $\boldsymbol{P}'_k$ is bounded, i.e., there exist positive scalars $\bar{\theta} > 0$ and $\underline{\theta} > 0$ such that:

$$\underline{\theta}\boldsymbol{I} \preceq \boldsymbol{P}'^{-1}_k \preceq \bar{\theta}\boldsymbol{I}. \tag{4.34}$$

**Assumption 4.2.** The matrix $\boldsymbol{A}_k$ is uniformly bounded and there exists $\boldsymbol{A}_k^{-1}$.

**Theorem 4.1.** *If*

$$\bar{\sigma}\left(\boldsymbol{\alpha}_k\right) \leq \gamma_1 = \frac{\underline{\sigma}\left(\boldsymbol{A}_k\right)}{\bar{\sigma}\left(\boldsymbol{A}_k\right)}\left(\frac{(1-\zeta)\underline{\sigma}\left(\boldsymbol{P}_k\right)}{\bar{\sigma}\left(\boldsymbol{A}_{1,k}\boldsymbol{P}'_k\boldsymbol{A}_{1,k}^T\right)}\right)^{\frac{1}{2}}, \tag{4.35}$$

*and*

$$\begin{aligned}\bar{\sigma}\left(\boldsymbol{\alpha}_k - \boldsymbol{I}\right) &\leq \gamma_2 \\ &= \frac{\underline{\sigma}\left(\boldsymbol{A}_k\right)}{\bar{\sigma}\left(\boldsymbol{A}_k\right)}\left(\frac{\underline{\sigma}\left(\boldsymbol{C}_k^T\right)\underline{\sigma}\left(\boldsymbol{C}_k\right)}{\bar{\sigma}\left(\boldsymbol{C}_k^T\right)\bar{\sigma}\left(\boldsymbol{C}_k\right)}\frac{\underline{\sigma}\left(\boldsymbol{R}_k\right)}{\bar{\sigma}\left(\boldsymbol{C}_k\boldsymbol{P}_k\boldsymbol{C}_k^T + \boldsymbol{R}_k\right)}\right)^{\frac{1}{2}},\end{aligned} \tag{4.36}$$

*where $0 < \zeta < 1$, then the proposed extended unknown input observer is locally asymptotically convergent.*

*Proof.* The main objective of further deliberations is to determine conditions under which the sequence $\{V_k\}_{k=1}^{\infty}$, defined by the Lyapunov candidate function

$$V_{k+1} = \boldsymbol{e}_{k+1}^T\boldsymbol{A}_{1,k+1}^{-T}[\boldsymbol{P}'_{k+1}]^{-1}\boldsymbol{A}_{1,k+1}^{-1}\boldsymbol{e}_{k+1}, \tag{4.37}$$

is a decreasing one. It should be pointed out that the Lyapunov function (4.37) involves a very restrictive assumption regarding an inverse of the matrix $\boldsymbol{A}_{1,k+1}^{-1}$. Indeed, from (4.15) and (4.7), (4.6) it is clear that the matrix $\boldsymbol{A}_{1,k+1}$ is singular when $\boldsymbol{E}_k \neq \boldsymbol{0}$. Thus, the convergence conditions can be formally obtained only when $\boldsymbol{E}_k = \boldsymbol{0}$. This means that the practical solution regarding the choice of the instrumental matrices $\boldsymbol{Q}_k$ and $\boldsymbol{R}_k$ is to be obtained for the case when $\boldsymbol{E}_k = \boldsymbol{0}$ and generalised to other cases, i.e., $\boldsymbol{E}_k \neq \boldsymbol{0}$.

First, let us define an alternative form of $\boldsymbol{K}_{1,k}$ and the inverse of $\boldsymbol{P}'_{k+1}$. Substituting (4.18) into (4.17) and then comparing it with (4.14), one can obtain

$$\boldsymbol{A}_{1,k+1}\boldsymbol{K}_{1,k+1}\boldsymbol{C}_k\boldsymbol{P}_k\boldsymbol{A}_{1,k+1}^T = \boldsymbol{K}_{1,k+1}\boldsymbol{C}_k\boldsymbol{P}_k. \tag{4.38}$$

Next, from (4.38), (4.18) and (4.16), the gain matrix becomes

$$\boldsymbol{K}_{1,k+1} = \boldsymbol{A}_{1,k+1}\boldsymbol{P}'_{k+1}\boldsymbol{C}_k^T\boldsymbol{R}_k^{-1}. \tag{4.39}$$

Similarly, from (4.38) and (4.18), the inverse of $\boldsymbol{P}'_{k+1}$ becomes

$$[\boldsymbol{P}'_{k+1}]^{-1} = \boldsymbol{P}_k^{-1} + \boldsymbol{C}_k^T\boldsymbol{R}_k^{-1}\boldsymbol{C}_k. \tag{4.40}$$

Substituting (4.33), and then (4.39) and (4.40) into (4.37), the Lyapunov candidate function is

$$\begin{aligned} V_{k+1} =& \boldsymbol{e}_k^T[\boldsymbol{A}_k^T\boldsymbol{\alpha}_k\boldsymbol{A}_k^{-T}\boldsymbol{P}_k^{-1}\boldsymbol{A}_k^{-1}\boldsymbol{\alpha}_k\boldsymbol{A}_k \\ &+ \boldsymbol{A}_k^T\boldsymbol{\alpha}_k\boldsymbol{A}_k^{-T}\boldsymbol{C}_k^T\boldsymbol{R}_k^{-1}\boldsymbol{C}_k\boldsymbol{A}_k^{-1}\boldsymbol{\alpha}_k\boldsymbol{A}_k \\ &- \boldsymbol{A}_k^T\boldsymbol{\alpha}_k\boldsymbol{A}_k^{-T}\boldsymbol{C}_k^T\boldsymbol{R}_k^{-1}\boldsymbol{C}_k - \boldsymbol{C}_k^T\boldsymbol{R}_k^{-1}\boldsymbol{C}_k\boldsymbol{A}_k^{-1}\boldsymbol{\alpha}_k\boldsymbol{A}_k \\ &+ \boldsymbol{C}_k^T\boldsymbol{R}_k^{-1}\boldsymbol{C}_k\boldsymbol{P}'_{k+1}\boldsymbol{C}_k^T\boldsymbol{R}_k^{-1}\boldsymbol{C}_k]\boldsymbol{e}_k. \end{aligned} \tag{4.41}$$

Let

$$\boldsymbol{G} = \boldsymbol{A}_k^{-1}\boldsymbol{\alpha}_k\boldsymbol{A}_k, \quad \boldsymbol{L} = \boldsymbol{C}_k^T\boldsymbol{R}_k^{-1}\boldsymbol{C}_k, \tag{4.42}$$

then

$$\boldsymbol{G}^T\boldsymbol{L}\boldsymbol{G} - \boldsymbol{G}^T\boldsymbol{L} - \boldsymbol{L}\boldsymbol{G} = \left[\boldsymbol{G}^T - \boldsymbol{I}\right]\boldsymbol{L}\left[\boldsymbol{G} - \boldsymbol{I}\right] - \boldsymbol{L}. \tag{4.43}$$

Using (4.43) and (4.16), the expression (4.41) becomes

$$\begin{aligned} V_{k+1} =& \boldsymbol{e}_k^T\left[\boldsymbol{A}_k^T\boldsymbol{\alpha}_k\boldsymbol{A}_k^{-T}\boldsymbol{P}_k^{-1}\boldsymbol{A}_k^{-1}\boldsymbol{\alpha}_k\boldsymbol{A}_k\right. \\ &+ \left[\boldsymbol{A}_k^T\boldsymbol{\alpha}_k\boldsymbol{A}_k^{-T} - \boldsymbol{I}\right]\boldsymbol{C}_k^T\boldsymbol{R}_k^{-1}\boldsymbol{C}_k\left[\boldsymbol{A}_k^{-1}\boldsymbol{\alpha}_k\boldsymbol{A}_k - \boldsymbol{I}\right] \\ &\left.-\boldsymbol{C}_k^T\boldsymbol{R}_k^{-1}\left[\boldsymbol{I} - \boldsymbol{C}_k\boldsymbol{P}_k\boldsymbol{C}_k^T\left[\boldsymbol{C}_k\boldsymbol{P}_k\boldsymbol{C}_k^T + \boldsymbol{R}_k\right]^{-1}\right]\right]\boldsymbol{e}_k. \end{aligned} \tag{4.44}$$

Using the identity in (4.44),

$$\boldsymbol{I} = \left[\boldsymbol{C}_k\boldsymbol{P}_k\boldsymbol{C}_k^T + \boldsymbol{R}_k\right]\left[\boldsymbol{C}_k\boldsymbol{P}_k\boldsymbol{C}_k^T + \boldsymbol{R}_k\right]^{-1}, \tag{4.45}$$

the Lyapunov candidate function can be written as

$$\begin{aligned} V_{k+1} =& \boldsymbol{e}_k^T\left[\boldsymbol{A}_k^T\boldsymbol{\alpha}_k\boldsymbol{A}_k^{-T}\boldsymbol{P}_k^{-1}\boldsymbol{A}_k^{-1}\boldsymbol{\alpha}_k\boldsymbol{A}_k\right. \\ &+ \left[\boldsymbol{A}_k^T\boldsymbol{\alpha}_k\boldsymbol{A}_k^{-T} - \boldsymbol{I}\right]\boldsymbol{C}_k^T\boldsymbol{R}_k^{-1}\boldsymbol{C}_k\left[\boldsymbol{A}_k^{-1}\boldsymbol{\alpha}_k\boldsymbol{A}_k - \boldsymbol{I}\right] \\ &\left.-\boldsymbol{C}_k^T\left[\boldsymbol{C}_k\boldsymbol{P}_k\boldsymbol{C}_k^T + \boldsymbol{R}_k\right]^{-1}\boldsymbol{C}_k\right]\boldsymbol{e}_k. \end{aligned} \tag{4.46}$$

The sequence $\{V_k\}_{k=1}^{\infty}$ is a decreasing one when there exists a scalar $\zeta$, $0 < \zeta < 1$, such that

$$V_{k+1} - (1-\zeta)V_k \le 0. \tag{4.47}$$

Using (4.46), the inequality (4.47) becomes

$$V_{k+1} - (1-\zeta)V_k = \boldsymbol{e}_k^T \boldsymbol{X}_k \boldsymbol{e}_k + \boldsymbol{e}_k^T \boldsymbol{Y}_k \boldsymbol{e}_k \le 0, \tag{4.48}$$

where

$$\boldsymbol{X}_k = \boldsymbol{A}_k^T \boldsymbol{\alpha}_k \boldsymbol{A}_k^{-T} \boldsymbol{P}_k^{-1} \boldsymbol{A}_k^{-1} \boldsymbol{\alpha}_k \boldsymbol{A}_k - (1-\zeta) \boldsymbol{A}_{1,k}^{-T} [\boldsymbol{P}'_k]^{-1} \boldsymbol{A}_{1,k}^{-1}, \tag{4.49}$$

$$\begin{aligned} \boldsymbol{Y}_k = & \left[ \boldsymbol{A}_k^T \boldsymbol{\alpha}_k \boldsymbol{A}_k^{-T} - \boldsymbol{I} \right] \boldsymbol{C}_k^T \boldsymbol{R}_k^{-1} \boldsymbol{C}_k \left[ \boldsymbol{A}_k^{-1} \boldsymbol{\alpha}_k \boldsymbol{A}_k - \boldsymbol{I} \right] \\ & - \boldsymbol{C}_k^T \left[ \boldsymbol{C}_k \boldsymbol{P}_k \boldsymbol{C}_k^T + \boldsymbol{R}_k \right]^{-1} \boldsymbol{C}_k. \end{aligned} \tag{4.50}$$

In order to satisfy (4.48), the matrices $\boldsymbol{X}_k$ and $\boldsymbol{Y}_k$ should be semi-negative defined. This is equivalent to

$$\bar{\sigma}\left( \boldsymbol{A}_k^T \boldsymbol{\alpha}_k \boldsymbol{A}_k^{-T} \boldsymbol{P}_k^{-1} \boldsymbol{A}_k^{-1} \boldsymbol{\alpha}_k \boldsymbol{A}_k \right) \le \underline{\sigma}\left( (1-\zeta) \boldsymbol{A}_{1,k}^{-T} [\boldsymbol{P}'_k]^{-1} \boldsymbol{A}_{1,k}^{-1} \right), \tag{4.51}$$

and

$$\begin{aligned} & \bar{\sigma}\left( \left[ \boldsymbol{A}_k^T \boldsymbol{\alpha}_k \boldsymbol{A}_k^{-T} - \boldsymbol{I} \right] \boldsymbol{C}_k^T \boldsymbol{R}_k^{-1} \boldsymbol{C}_k \left[ \boldsymbol{A}_k^{-1} \boldsymbol{\alpha}_k \boldsymbol{A}_k - \boldsymbol{I} \right] \right) \\ & \le \underline{\sigma}\left( \boldsymbol{C}_k^T \left[ \boldsymbol{C}_k \boldsymbol{P}_k \boldsymbol{C}_k^T + \boldsymbol{R}_k \right]^{-1} \boldsymbol{C}_k \right). \end{aligned} \tag{4.52}$$

The inequalities (4.51) and (4.52) determine the bounds of the diagonal matrix $\boldsymbol{\alpha}_k$, for which the condition (4.48) is satisfied. The objective of further deliberations is to obtain a more convenient form of the above bounds. Using the fact that

$$\begin{aligned} \bar{\sigma}\left( \boldsymbol{A}_k^T \boldsymbol{\alpha}_k \boldsymbol{A}_k^{-T} \boldsymbol{P}_k^{-1} \boldsymbol{A}_k^{-1} \boldsymbol{\alpha}_k \boldsymbol{A}_k \right) & \le \bar{\sigma}^2(\boldsymbol{A}_k) \bar{\sigma}^2(\boldsymbol{A}_k^{-1}) \bar{\sigma}^2(\boldsymbol{\alpha}_k) \bar{\sigma}(\boldsymbol{P}_k^{-1}) \\ & = \frac{\bar{\sigma}^2(\boldsymbol{A}_k) \bar{\sigma}^2(\boldsymbol{\alpha}_k)}{\underline{\sigma}^2(\boldsymbol{A}_k) \underline{\sigma}(\boldsymbol{P}_k)}, \end{aligned} \tag{4.53}$$

the expression (4.51) gives (4.35). Similarly, using

$$\begin{aligned} \bar{\sigma}\left( \left[ \boldsymbol{A}_k^T \boldsymbol{\alpha}_k \boldsymbol{A}_k^{-T} - \boldsymbol{I} \right] \right) & = \bar{\sigma}\left( \boldsymbol{A}_k^T \left[ \boldsymbol{\alpha}_k - \boldsymbol{I} \right] \boldsymbol{A}_k^{-T} \right) \\ & \le \frac{\bar{\sigma}(\boldsymbol{A}_k)}{\underline{\sigma}(\boldsymbol{A}_k)} \bar{\sigma}(\boldsymbol{\alpha}_k - \boldsymbol{I}), \end{aligned} \tag{4.54}$$

and then

$$\begin{aligned} & \bar{\sigma}\left( \left[ \boldsymbol{A}_k^T \boldsymbol{\alpha}_k \boldsymbol{A}_k^{-T} - \boldsymbol{I} \right] \boldsymbol{C}_k^T \boldsymbol{R}_k^{-1} \boldsymbol{C}_k \left[ \boldsymbol{A}_k^{-1} \boldsymbol{\alpha}_k \boldsymbol{A}_k - \boldsymbol{I} \right] \right) \\ & \le \frac{\bar{\sigma}^2(\boldsymbol{A}_k)}{\underline{\sigma}^2(\boldsymbol{A}_k)} \frac{\bar{\sigma}\left( \boldsymbol{C}_k^T \right) \bar{\sigma}(\boldsymbol{C}_k)}{\underline{\sigma}(\boldsymbol{R}_k)} \bar{\sigma}^2(\boldsymbol{\alpha}_k - \boldsymbol{I}) \end{aligned} \tag{4.55}$$

and

$$\underline{\sigma}\left(\boldsymbol{C}_k^T\left[\boldsymbol{C}_k\boldsymbol{P}_k\boldsymbol{C}_k^T+\boldsymbol{R}_k\right]^{-1}\boldsymbol{C}_k\right)\geq\frac{\underline{\sigma}\left(\boldsymbol{C}_k^T\right)\underline{\sigma}\left(\boldsymbol{C}_k\right)}{\bar{\sigma}\left(\boldsymbol{C}_k\boldsymbol{P}_k\boldsymbol{C}_k^T+\boldsymbol{R}_k\right)}, \tag{4.56}$$

the expression (4.52) gives (4.36).

Thus, if the conditions (4.35) and (4.36) are satisfied, then $\{V_k\}_{k=1}^{\infty}$ is a decreasing sequence and hence, under Assumption 4.1, the proposed observer is locally asymptotically convergent.

Bearing in mind that $\boldsymbol{\alpha}_k$ is a diagonal matrix, the inequalities (4.35)–(4.36) can be expressed as

$$\max_{i=1,\ldots,n}|\alpha_{i,k}|\leq\gamma_1 \quad \text{and} \quad \max_{i=1,\ldots,n}|\alpha_{i,k}-1|\leq\gamma_2. \tag{4.57}$$

Since

$$\boldsymbol{P}_k=\boldsymbol{A}_{1,k}\boldsymbol{P}_k'\boldsymbol{A}_{1,k}^T+\boldsymbol{T}_k\boldsymbol{Q}_{k-1}\boldsymbol{T}_k^T+\boldsymbol{H}_k\boldsymbol{R}_k\boldsymbol{H}_k^T, \tag{4.58}$$

it is clear that an appropriate selection of the instrumental matrices $\boldsymbol{Q}_{k-1}$ and $\boldsymbol{R}_k$ may enlarge the bounds $\gamma_1$ and $\gamma_2$ and, consequently, the domain of attraction. Indeed, if the conditions (4.57) are satisfied, then $\hat{\boldsymbol{x}}_k$ converges to $\boldsymbol{x}_k$.

Unfortunately, analytical derivation of the matrices $\boldsymbol{Q}_{k-1}$ and $\boldsymbol{R}_k$ seems to be an extremely difficult problem. However, it is possible to set the above matrices as follows: $\boldsymbol{Q}_{k-1}=\beta_1\boldsymbol{I}$, $\boldsymbol{R}_k=\beta_1\boldsymbol{I}$, with $\beta_1$ and $\beta_1$ large enough. On the other hand, it is well known that the convergence rate of such an EKF-like approach can be increased by an appropriate selection of the covariance matrices $\boldsymbol{Q}_{k-1}$ and $\boldsymbol{R}_k$, i.e., the more accurate (near "true" values) the covariance matrices, the better the convergence rate. This means that in the deterministic case ($\boldsymbol{w}_k=\boldsymbol{0}$ and $\boldsymbol{v}_k=\boldsymbol{0}$), both matrices should be zero ones. Unfortunately, such an approach usually leads to the divergence of the observer as well as other computational problems. To tackle this, a compromise between the convergence and the convergence rate should be established. This can easily be done by setting the instrumental matrices as

$$\boldsymbol{Q}_{k-1}=\beta_1\boldsymbol{\varepsilon}_{k-1}^T\boldsymbol{\varepsilon}_{k-1}\boldsymbol{I}+\delta_1\boldsymbol{I}, \quad \boldsymbol{R}_k=\beta_2\boldsymbol{\varepsilon}_k^T\boldsymbol{\varepsilon}_k\boldsymbol{I}+\delta_2\boldsymbol{I}, \tag{4.59}$$

with $\beta_1$, $\beta_2$ large enough, and $\delta_1$, $\delta_2$ small enough.

### 4.1.3 Illustrative Example

The purpose of this section is to show the reliability and effectiveness of the proposed EUIO. The numerical example considered here is a fifth-order two-phase non-linear model of an induction motor, which has already been the subject of a large number of various control design applications (see [19] and the references therein). The complete discrete-time model in a stator-fixed $(a,b)$ reference frame is

$$x_{1,k+1} = x_{1,k} + h\left(-\gamma x_{1k} + \frac{K}{T_r}x_{3k} + Kpx_{5k}x_{4k} + \frac{1}{\sigma L_s}u_{1k}\right), \tag{4.60}$$

$$x_{2,k+1} = x_{2,k} + h\left(-\gamma x_{2k} - Kpx_{5k}x_{3k} + \frac{K}{T_r}x_{4k} + \frac{1}{\sigma L_s}u_{2k}\right), \tag{4.61}$$

$$x_{3,k+1} = x_{3,k} + h\left(\frac{M}{T_r}x_{1k} - \frac{1}{T_r}x_{3k} - px_{5k}x_{4k}\right), \tag{4.62}$$

$$x_{4,k+1} = x_{4,k} + h\left(\frac{M}{T_r}x_{2k} + px_{5k}x_{3k} - \frac{1}{T_r}x_{4k}\right), \tag{4.63}$$

$$x_{5,k+1} = x_{5,k} + h\left(\frac{pM}{JL_r}(x_{3k}x_{2k} - x_{4k}x_{1k}) - \frac{T_L}{J}\right), \tag{4.64}$$

$$y_{1,k+1} = x_{1,k+1}, \; y_{2,k+1} = x_{2,k+1}, \tag{4.65}$$

while $\boldsymbol{x}_k = [x_{1,k}, \ldots, x_{n,k}]^T = [i_{\text{sak}}, i_{\text{sbk}}, \psi_{\text{rak}}, \psi_{\text{rbk}}, \omega_k]^T$ represents the currents, the rotor fluxes, and the angular speed, respectively, while $\boldsymbol{u}_k = [u_{\text{sak}}, u_{\text{sbk}}]^T$ is the stator voltage control vector, $p$ is the number of the pairs of poles, and $T_L$ is the load torque. The rotor time constant $T_r$ and the remaining parameters are defined as

$$T_r = \frac{L_r}{R_r}, \; \sigma = 1 - \frac{M^2}{L_s L_r}, \; K = \frac{M}{\sigma L_s L_r}, \; \gamma = \frac{R_s}{\sigma L_s} + \frac{R_r M^2}{\sigma L_s L_r^2}, \tag{4.66}$$

while $R_s$, $R_r$ and $L_s$, $L_r$ are stator and rotor per phase resistances and inductances, respectively, and $J$ is the rotor moment inertia.

The numerical values of the above parameters are as follows: $R_s = 0.18\ \Omega$, $R_r = 0.15\ \Omega$, $M = 0.068$ H, $L_s = 0.0699$ H, $L_r = 0.0699$ H, $J = 0.0586$ kgm$^2$, $T_L = 10$ Nm, $p = 1$, and $h = 0.1$ ms. The input signals are

$$u_{1,k} = 350\cos(0.03k), \quad u_{2,k} = 300\sin(0.03k). \tag{4.67}$$

The initial conditions for the system and the observer are $\boldsymbol{x}_k = \boldsymbol{0}$ and $\hat{\boldsymbol{x}}_k = [200, 200, 50, 50, 300]^T$, and $\boldsymbol{P}_0 = 10^3\boldsymbol{I}$.

Moreover, the following two cases concerning the selection of $\boldsymbol{Q}_{k-1}$ and $\boldsymbol{R}_k$ were considered:

Case 1: Classic approach (constant values), i.e., $\boldsymbol{Q}_{k-1} = 0.1\boldsymbol{I}$, $\boldsymbol{R}_k = 0.1\boldsymbol{I}$,
Case 2: Selection according to (4.59), i.e.,

$$\boldsymbol{Q}_{k-1} = 10^3\boldsymbol{\varepsilon}_{k-1}^T\boldsymbol{\varepsilon}_{k-1}\boldsymbol{I} + 0.01\boldsymbol{I},$$
$$\boldsymbol{R}_k = 10\boldsymbol{\varepsilon}_k^T\boldsymbol{\varepsilon}_k\boldsymbol{I} + 0.01\boldsymbol{I}. \tag{4.68}$$

The results shown in Fig. 4.1 confirm the relevance of appropriate selection of the instrumental matrices. Indeed, as can be seen, the proposed approach is superior to the classic technique of selecting the instrumental matrices $\boldsymbol{Q}_{k-1}$ and $\boldsymbol{R}_k$.

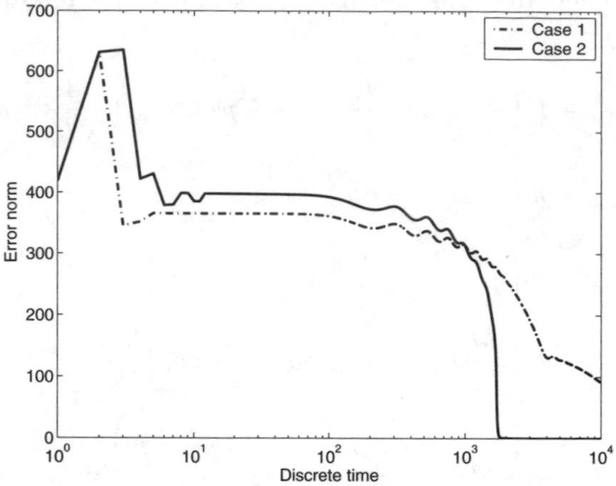


**Fig. 4.1.** State estimation error norm $\|e_k\|_2$ for Case 1 and Case 2

Apart from the relatively good results presented in Fig. 4.1, it can be shown that the application of soft computing techniques [96, 187] makes it possible to increase the convergence rate further. The details regarding such an approach are presented in Section 6.2.

## 4.2 Extended Unknown Input Observer Revisited

The first objective of this section is to present two different approaches that can be used for state estimation of a non-linear discrete-time system described by

$$\boldsymbol{x}_{k+1} = \boldsymbol{g}(\boldsymbol{x}_k) + \boldsymbol{h}(\boldsymbol{u}_k) + \boldsymbol{E}_k\boldsymbol{d}_k, \tag{4.69}$$

$$\boldsymbol{y}_{k+1} = \boldsymbol{C}_{k+1}\boldsymbol{x}_{k+1}. \tag{4.70}$$

The second and main objective is to show that the schemes being presented are equivalent. This property is then employed in the subsequent part of this work to form a novel UIO structure and to prove its convergence under less restrictive assumptions than those used in [187]. Finally, it should be pointed out that the research results portrayed in this section were originally presented in [189].

In the existing approaches, the unknown input is usually treated in two different ways. The first one (see, e.g., [27]) consists in introducing an additional matrix into the state estimation equation, which is then used for decoupling the effect of the unknown input on the state estimation error (and, consequently, on the residual signal). In the second approach (see, e.g., [91]), the system with an unknown input is suitably transformed into a system without it. In both cases, the necessary condition for the existence of a solution to the unknown input decoupling problem is (see Section 4.1):

$$\mathrm{rank}(\boldsymbol{C}_{k+1}\boldsymbol{E}_k) = \mathrm{rank}(\boldsymbol{E}_k) = q, \tag{4.71}$$

(see [27, p. 72, Lemma 3.1] for a comprehensive explanation). If the condition (4.71) is satisfied, then it is possible to calculate $\boldsymbol{H}_{k+1} = (\boldsymbol{C}_{k+1}\boldsymbol{E}_k)^+ = \left[(\boldsymbol{C}_{k+1}\boldsymbol{E}_k)^T\boldsymbol{C}_{k+1}\boldsymbol{E}_k\right]^{-1}(\boldsymbol{C}_{k+1}\boldsymbol{E}_k)^T$, where $(\cdot)^+$ stands for the pseudo-inverse of its argument. Thus, by multiplying (4.70) by $\boldsymbol{H}_{k+1}$ and then inserting (4.69), it is straightforward to show that

$$\boldsymbol{d}_k = \boldsymbol{H}_{k+1}\left[\boldsymbol{y}_{k+1} - \boldsymbol{C}_{k+1}\left[\boldsymbol{g}(\boldsymbol{x}_k) + \boldsymbol{h}(\boldsymbol{u}_k)\right]\right]. \tag{4.72}$$

Substituting (4.72) into (4.69) gives

$$\boldsymbol{x}_{k+1} = \bar{\boldsymbol{g}}(\boldsymbol{x}_k) + \bar{\boldsymbol{h}}(\boldsymbol{u}_k) + \bar{\boldsymbol{E}}_k\boldsymbol{y}_{k+1}, \tag{4.73}$$

where

$$\bar{\boldsymbol{g}}(\cdot) = \bar{\boldsymbol{G}}_k\boldsymbol{g}(\cdot), \ \bar{\boldsymbol{h}}(\cdot) = \bar{\boldsymbol{G}}_k\boldsymbol{h}(\cdot), \ \bar{\boldsymbol{G}}_k = \boldsymbol{I} - \boldsymbol{E}_k\boldsymbol{H}_{k+1}\boldsymbol{C}_{k+1}, \ \bar{\boldsymbol{E}}_k = \boldsymbol{E}_k\boldsymbol{H}_{k+1}.$$

Thus, the unknown input observer for (4.69)–(4.70) is given as follows:

$$\hat{\boldsymbol{x}}_{k+1} = \bar{\boldsymbol{g}}(\hat{\boldsymbol{x}}_k) + \bar{\boldsymbol{h}}(\boldsymbol{u}_k) + \bar{\boldsymbol{E}}_k\boldsymbol{y}_{k+1} + \boldsymbol{K}_{k+1}(\boldsymbol{y}_k - \boldsymbol{C}_k\hat{\boldsymbol{x}}_k). \tag{4.74}$$

Now, let us consider the first of the above-mentioned approaches, which can be used for designing the UIO [27]. This approach was used in Section 4.1. For notational simplicity, let us start with the UIO for linear discrete-time systems:

$$\begin{aligned}\boldsymbol{x}_{k+1} &= \boldsymbol{A}_k\boldsymbol{x}_k + \boldsymbol{B}_k\boldsymbol{u}_k + \boldsymbol{E}_k\boldsymbol{d}_k,\\ \boldsymbol{y}_{k+1} &= \boldsymbol{C}_{k+1}\boldsymbol{x}_{k+1},\end{aligned} \tag{4.75}$$

which can be described as follows [27] (see also Section 2.3):

$$\boldsymbol{s}_{k+1} = \boldsymbol{F}_{k+1}\boldsymbol{s}_k + \boldsymbol{T}_{k+1}\boldsymbol{B}_k\boldsymbol{u}_k + \boldsymbol{K}_{1,k+1}\boldsymbol{y}_k, \tag{4.76}$$

$$\hat{\boldsymbol{x}}_{k+1} = \boldsymbol{s}_{k+1} + \boldsymbol{H}_{1,k+1}\boldsymbol{y}_{k+1}, \tag{4.77}$$

where

$$\boldsymbol{K}_{1,k+1} = \boldsymbol{K}_{k+1} + \boldsymbol{K}_{2,k+1}, \tag{4.78}$$

$$\boldsymbol{E}_k = \boldsymbol{H}_{1,k+1}\boldsymbol{C}_{k+1}\boldsymbol{E}_k, \tag{4.79}$$

$$\boldsymbol{T}_{k+1} = \boldsymbol{I} - \boldsymbol{H}_{1,k+1}\boldsymbol{C}_{k+1}, \tag{4.80}$$

$$\boldsymbol{F}_{k+1} = \boldsymbol{T}_{k+1}\boldsymbol{A}_k - \boldsymbol{K}_{k+1}\boldsymbol{C}_k, \tag{4.81}$$

$$\boldsymbol{K}_{2,k+1} = \boldsymbol{F}_{k+1}\boldsymbol{H}_{1,k}. \tag{4.82}$$

By substituting (4.77) into (4.76) and then using (4.80), (4.81) and (4.82), it can be shown that

$$\begin{aligned}\hat{\boldsymbol{x}}_{k+1} =& \boldsymbol{A}_k\hat{\boldsymbol{x}}_k + \boldsymbol{B}_k\boldsymbol{u}_k - \boldsymbol{H}_{1,k+1}\boldsymbol{C}_{k+1}[\boldsymbol{A}_k\hat{\boldsymbol{x}}_k + \boldsymbol{B}_k\boldsymbol{u}_k] +\\ &- \boldsymbol{K}_{k+1}\boldsymbol{C}_k\hat{\boldsymbol{x}}_k - \boldsymbol{F}_{k+1}\boldsymbol{H}_{1,k+1}\boldsymbol{y}_k +\\ &+ [\boldsymbol{K}_{k+1} + \boldsymbol{F}_{k+1}\boldsymbol{H}_{1,k+1}]\boldsymbol{y}_k + \boldsymbol{H}_{1,k+1}\boldsymbol{y}_{k+1},\end{aligned} \tag{4.83}$$

or, equivalently,

$$\hat{\boldsymbol{x}}_{k+1} = \hat{\boldsymbol{x}}_{k+1/k} + \boldsymbol{H}_{1,k+1}(\boldsymbol{y}_{k+1} - \boldsymbol{C}_{k+1}\hat{\boldsymbol{x}}_{k+1/k}) + \boldsymbol{K}_{k+1}(\boldsymbol{y}_k - \boldsymbol{C}_k\hat{\boldsymbol{x}}_k), \quad (4.84)$$

and

$$\hat{\boldsymbol{x}}_{k+1/k} = \boldsymbol{A}_k\hat{\boldsymbol{x}}_k + \boldsymbol{B}_k\boldsymbol{u}_k. \quad (4.85)$$

Substituting the solution of (4.79), i.e., $\boldsymbol{H}_{1,k+1} = \boldsymbol{E}_k\boldsymbol{H}_{k+1}$ into (4.84) yields

$$\begin{aligned}\hat{\boldsymbol{x}}_{k+1} =& [\boldsymbol{I} - \boldsymbol{E}_k\boldsymbol{H}_{k+1}\boldsymbol{C}_{k+1}]\hat{\boldsymbol{x}}_{k+1/k} + \\ &+ \boldsymbol{E}_k\boldsymbol{H}_{k+1}\boldsymbol{y}_{k+1} + \boldsymbol{K}_{k+1}(\boldsymbol{y}_k - \boldsymbol{C}_k\hat{\boldsymbol{x}}_k). \end{aligned} \quad (4.86)$$

Thus, in order to use (4.86) for (4.69)–(4.70) it is necessary to replace (4.85) by

$$\hat{\boldsymbol{x}}_{k+1/k} = \boldsymbol{g}\left(\hat{\boldsymbol{x}}_k\right) + \boldsymbol{h}(\boldsymbol{u}_k). \quad (4.87)$$

Finally, by substituting (4.87) into (4.86) and then comparing it with (4.74) it can be seen that the observer structures being considered are identical. On the other hand, it should be clearly pointed out that they were designed in significantly different ways. Following the above results, it is clear that the observer proposed in Section 4.1 can be designed in two alternative ways.

As was mentioned in Section 4.1, the convergence conditions exposed in Theorem 4.1 were developed under very restrictive assumptions. It seems that one possible approach to overcome the above-mentioned restrictive assumptions is to use a different strategy for convergence analysis. An approach alternative to the one presented in Section 4.1 and in [19, 187] was proposed by Guo and Zhu [72].

For the sake of simplicity, let us assume that $\boldsymbol{E}_k = \boldsymbol{0}$. Instead of using (4.32), Guo and Zhu proposed the following approach (as was the case in [72])

$$\boldsymbol{e}_{k+1/k} = \boldsymbol{g}\left(\boldsymbol{x}_k\right) - \boldsymbol{g}\left(\hat{\boldsymbol{x}}_k\right) = \boldsymbol{A}_k\boldsymbol{e}_k, \quad (4.88)$$

where

$$\boldsymbol{A}_k = \left.\frac{\partial \boldsymbol{g}\left(\boldsymbol{x}\right)}{\partial \boldsymbol{x}}\right|_{\boldsymbol{x}=\hat{\boldsymbol{x}}_k+\boldsymbol{\Delta}_k}. \quad (4.89)$$

where $\boldsymbol{\Delta}_k \in \mathbb{R}^n$, and $\hat{\boldsymbol{x}}_k + \boldsymbol{\Delta}_k$ is between $\boldsymbol{x}_k$ and $\hat{\boldsymbol{x}}_k$. This means that there exists a scalar $\lambda \in [0, 1]$ such that $\hat{\boldsymbol{x}}_k + \boldsymbol{\Delta}_k = \lambda\boldsymbol{x}_k + (1-\lambda)\hat{\boldsymbol{x}}_k$.

Unfortunately, in general, the above approach is incorrect as can be demonstrated by a counterexample. Namely, let us consider the following structure of $\boldsymbol{g}\left(\boldsymbol{x}\right)$:

$$\boldsymbol{g}\left(\boldsymbol{x}\right) = \left[x_1^2,\ e^{x_1+x_2}\right]^T, \quad (4.90)$$

with $\boldsymbol{x}_k = \boldsymbol{0}$ and $\hat{\boldsymbol{x}}_k = \boldsymbol{1}$ for which

$$\boldsymbol{g}\left(\boldsymbol{x}_k\right) = \boldsymbol{g}\left(\boldsymbol{0}\right) = [0,\ 1]^T, \quad \text{and} \quad \boldsymbol{g}\left(\hat{\boldsymbol{x}}_k\right) = \boldsymbol{g}\left(\boldsymbol{1}\right) = [1, e^2]^T. \quad (4.91)$$

Thus

$$\boldsymbol{g}\left(\boldsymbol{x}_k\right) - \boldsymbol{g}\left(\hat{\boldsymbol{x}}_k\right) = [-1,\ 1-e^2]^T, \quad (4.92)$$

and

$$\boldsymbol{A}_k\boldsymbol{e}_k = [-2(1-\lambda), -2e^{2(1-\lambda)}]^T. \tag{4.93}$$

This means that there should exist a scalar $\lambda \in [0,1]$ such that

$$[-1,\ 1-e^2]^T = [-2(1-\lambda), -2e^{2(1-\lambda)}]^T. \tag{4.94}$$

From the first equation of the above set of equations it is clear that $\lambda = \frac{1}{2}$, which is not valid for the second equation, i.e., $1-e^2 \neq -2e$.

This counterexample clearly shows that the observer convergence conditions described in [72] are invalid.

### 4.2.1 Convergence Analysis and Design Principles

Taking into account all the above-exposed difficulties, the main objective of this section is to propose an alternative structure of the EUIO and to derive its convergence conditions.

As has already been shown, the state equation (4.69) can be transformed into (4.73), but instead of using the observer structure (4.74) it is proposed to use its minor modification that can be given as

$$\hat{\boldsymbol{x}}_{k+1} = \hat{\boldsymbol{x}}_{k+1/k} + \boldsymbol{K}_{k+1}(\boldsymbol{y}_{k+1} - \boldsymbol{C}_{k+1}\hat{\boldsymbol{x}}_{k+1/k}), \tag{4.95}$$

where

$$\hat{\boldsymbol{x}}_{k+1/k} = \bar{\boldsymbol{g}}\left(\hat{\boldsymbol{x}}_k\right) + \bar{\boldsymbol{h}}\left(\boldsymbol{u}_k\right) + \bar{\boldsymbol{E}}_k\boldsymbol{y}_{k+1}. \tag{4.96}$$

As a consequence, the algorithm used for state estimation of (4.69)–(4.70) is given as follows:

$$\hat{\boldsymbol{x}}_{k+1/k} = \bar{\boldsymbol{g}}\left(\hat{\boldsymbol{x}}_k\right) + \bar{\boldsymbol{h}}\left(\boldsymbol{u}_k\right) + \bar{\boldsymbol{E}}_k\boldsymbol{y}_{k+1}, \tag{4.97}$$

$$\boldsymbol{P}_{k+1/k} = \bar{\boldsymbol{A}}_k\boldsymbol{P}_k\bar{\boldsymbol{A}}_k^{\ T} + \boldsymbol{Q}_k, \tag{4.98}$$

$$\boldsymbol{K}_{k+1} = \boldsymbol{P}_{k+1/k}\boldsymbol{C}_{k+1}^T\left(\boldsymbol{C}_{k+1}\boldsymbol{P}_{k+1/k}\boldsymbol{C}_{k+1}^T + \boldsymbol{R}_{k+1}\right)^{-1}, \tag{4.99}$$

$$\hat{\boldsymbol{x}}_{k+1} = \hat{\boldsymbol{x}}_{k+1/k} + \boldsymbol{K}_{k+1}(\boldsymbol{y}_{k+1} - \boldsymbol{C}_{k+1}\hat{\boldsymbol{x}}_{k+1/k}), \tag{4.100}$$

$$\boldsymbol{P}_{k+1} = \left[\boldsymbol{I} - \boldsymbol{K}_{k+1}\boldsymbol{C}_{k+1}\right]\boldsymbol{P}_{k+1/k}, \tag{4.101}$$

with

$$\bar{\boldsymbol{A}}_k = \left.\frac{\partial\bar{\boldsymbol{g}}\left(\boldsymbol{x}_k\right)}{\partial\boldsymbol{x}_k}\right|_{\boldsymbol{x}_k=\hat{\boldsymbol{x}}_k} = \bar{\boldsymbol{G}}_k\left.\frac{\partial\boldsymbol{g}\left(\boldsymbol{x}_k\right)}{\partial\boldsymbol{x}_k}\right|_{\boldsymbol{x}_k=\hat{\boldsymbol{x}}_k} = \bar{\boldsymbol{G}}_k\boldsymbol{A}_k. \tag{4.102}$$

The main objective of the subsequent part of this section is to present the convergence conditions of the proposed EUIO. In particular, the main aim is to show that the convergence of the EUIO strongly depends on the instrumental matrices $\boldsymbol{Q}_k$ and $\boldsymbol{R}_k$. Moreover, the fault-free mode is assumed, i.e., $\boldsymbol{f}_k = \boldsymbol{0}$.

Using (4.100), the state estimation error can be given as

$$\boldsymbol{e}_{k+1} = \boldsymbol{x}_{k+1} - \hat{\boldsymbol{x}}_{k+1} = \left[\boldsymbol{I} - \boldsymbol{K}_{k+1}\boldsymbol{C}_{k+1}\right]\boldsymbol{e}_{k+1/k}, \tag{4.103}$$

while

$$e_{k+1/k} = x_{k+1} - \hat{x}_{k+1/k} = \bar{g}(x_k) - \bar{g}(\hat{x}_k) = \alpha_k \bar{A}_k e_k, \tag{4.104}$$

where $\alpha_k = \mathrm{diag}(\alpha_{1,k}, \ldots, \alpha_{n,k})$ is an unknown diagonal matrix. Thus, using (4.104), the equation (4.103) becomes

$$e_{k+1} = [I - K_{k+1} C_{k+1}] \alpha_k \bar{A}_k e_k. \tag{4.105}$$

It is clear from (4.104) that $\alpha_k$ represents the linearisation error. This means that the convergence of the proposed observer is strongly related to the admissible bounds of the diagonal elements of $\alpha_k$. Thus, the main objective of further deliberations is to show that these bounds can be controlled with the use of the instrumental matrices $Q_k$ and $R_k$.

First, let us start with the convergence conditions, which require the following assumptions:

**Assumption 4.3.** Following [19], it is assumed that the system given by (4.73) and (4.70) is locally uniformly rank observable. This guaranees (see [19] and the references therein) that the matrix $P_k$ is bounded, i.e., there exist positive scalars $\bar{\theta} > 0$ and $\underline{\theta} > 0$ such that

$$\underline{\theta} I \preceq P_k^{-1} \preceq \bar{\theta} I. \tag{4.106}$$

**Assumption 4.4.** The matrix $A_k$ is uniformly bounded and there exists $A_k^{-1}$.

Moreover, let us define

$$\bar{\alpha}_k = \max_{j=1,\ldots,n} |\alpha_{j,k}|, \quad \underline{\alpha}_k = \min_{j=1,\ldots,n} |\alpha_{j,k}|. \tag{4.107}$$

**Theorem 4.2.** *If*

$$\bar{\alpha}_k \leq \left( \underline{\alpha}_k^2 \frac{\underline{\sigma}\left(\bar{A}_k\right)^2 \underline{\sigma}\left(C_{k+1}\right)^2 \underline{\sigma}\left(\bar{A}_k P_k \bar{A}_k^T + Q_k\right)}{\bar{\sigma}\left(C_{k+1} P_{k+1/k} C_{k+1}^T + R_{k+1}\right)} + \right.$$
$$\left. + \frac{(1-\zeta)\underline{\sigma}\left(\bar{A}_k P_k \bar{A}_k^T + Q_k\right)}{\bar{\sigma}\left(\bar{A}_k\right)^2 \bar{\sigma}\left(P_k\right)} \right)^{\frac{1}{2}}, \tag{4.108}$$

*where* $0 < \zeta < 1$, *then the proposed extended unknown input observer is locally asymptotically convergent.*

*Proof.* The main objective of further deliberations is to determine conditions for which the sequence $\{V_k\}_{k=1}^{\infty}$, defined by the Lyapunov candidate function

$$V_{k+1} = e_{k+1}^T P_{k+1}^{-1} e_{k+1}, \tag{4.109}$$

is a decreasing one. Substituting (4.105) into (4.109) gives

$$V_{k+1} = \boldsymbol{e}_k^T \bar{\boldsymbol{A}}_k{}^T \boldsymbol{\alpha}_k \cdot \\ \cdot \left[\boldsymbol{I} - \boldsymbol{C}_{k+1}^T \boldsymbol{K}_{k+1}^T\right] \boldsymbol{P}_{k+1}^{-1} \left[\boldsymbol{I} - \boldsymbol{K}_{k+1}\boldsymbol{C}_{k+1}\right] \boldsymbol{\alpha}_k \bar{\boldsymbol{A}}_k \boldsymbol{e}_k. \tag{4.110}$$

Using (4.101), it can be shown that

$$\left[\boldsymbol{I} - \boldsymbol{C}_{k+1}^T \boldsymbol{K}_{k+1}^T\right] = \boldsymbol{P}_{k+1/k}^{-1} \boldsymbol{P}_{k+1}. \tag{4.111}$$

Inserting (4.99) into $[\boldsymbol{I} - \boldsymbol{K}_{k+1}\boldsymbol{C}_{k+1}]$ yields

$$[\boldsymbol{I} - \boldsymbol{K}_{k+1}\boldsymbol{C}_{k+1}] = \boldsymbol{P}_{k+1/k} \cdot \\ \cdot \left[\boldsymbol{P}_{k+1/k}^{-1} - \boldsymbol{C}_{k+1}^T \left(\boldsymbol{C}_{k+1}\boldsymbol{P}_{k+1/k}\boldsymbol{C}_{k+1}^T + \boldsymbol{R}_{k+1}\right)^{-1} \boldsymbol{C}_{k+1}\right]. \tag{4.112}$$

Substituting (4.111) and (4.112) into (4.110) gives

$$V_{k+1} = \boldsymbol{e}_k^T \bar{\boldsymbol{A}}_k{}^T \boldsymbol{\alpha}_k \cdot \\ \cdot \left[\boldsymbol{P}_{k+1/k}^{-1} - \boldsymbol{C}_{k+1}^T \left(\boldsymbol{C}_{k+1}\boldsymbol{P}_{k+1/k}\boldsymbol{C}_{k+1}^T + \boldsymbol{R}_{k+1}\right)^{-1} \boldsymbol{C}_{k+1}\right] \boldsymbol{\alpha}_k \bar{\boldsymbol{A}}_k \boldsymbol{e}_k. \tag{4.113}$$

The sequence $\{V_k\}_{k=1}^{\infty}$ is decreasing when there exists a scalar $\zeta$, $0 < \zeta < 1$, such that

$$V_{k+1} - (1-\zeta)V_k \leq 0. \tag{4.114}$$

Using (4.109) and (4.113), the inequality (4.114) can be written as

$$\boldsymbol{e}_k^T \left[\bar{\boldsymbol{A}}_k{}^T \boldsymbol{\alpha}_k \left[\boldsymbol{P}_{k+1/k}^{-1} - \boldsymbol{C}_{k+1}^T \left(\boldsymbol{C}_{k+1}\boldsymbol{P}_{k+1/k}\boldsymbol{C}_{k+1}^T + \boldsymbol{R}_{k+1}\right)^{-1} \boldsymbol{C}_{k+1}\right] \boldsymbol{\alpha}_k \bar{\boldsymbol{A}}_k \\ -(1-\zeta)\boldsymbol{P}_k^{-1}\right] \boldsymbol{e}_k \leq 0. \tag{4.115}$$

Using the bounds of the Rayleigh quotient for $\boldsymbol{X} \succeq 0$, i.e., $\underline{\sigma}(\boldsymbol{X}) \leq \frac{\boldsymbol{e}_k^T \boldsymbol{X} \boldsymbol{e}_k}{\boldsymbol{e}_k^T \boldsymbol{e}_k} \leq \bar{\sigma}(\boldsymbol{X})$, the inequality (4.115) can be transformed into the following form:

$$\bar{\sigma}\left(\bar{\boldsymbol{A}}_k{}^T \boldsymbol{\alpha}_k \boldsymbol{P}_{k+1/k}^{-1} \boldsymbol{\alpha}_k \bar{\boldsymbol{A}}_k\right) + \\ - \underline{\sigma}\left(\bar{\boldsymbol{A}}_k{}^T \boldsymbol{\alpha}_k \boldsymbol{C}_{k+1}^T \left(\boldsymbol{C}_{k+1}\boldsymbol{P}_{k+1/k}\boldsymbol{C}_{k+1}^T + \boldsymbol{R}_{k+1}\right)^{-1} \boldsymbol{C}_{k+1} \boldsymbol{\alpha}_k \bar{\boldsymbol{A}}_k\right) + \\ - (1-\zeta)\underline{\sigma}\left(\boldsymbol{P}_k^{-1}\right) \leq 0. \tag{4.116}$$

It is straightforward to show that

$$\bar{\sigma}\left(\bar{\boldsymbol{A}}_k{}^T \boldsymbol{\alpha}_k \boldsymbol{P}_{k+1/k}^{-1} \boldsymbol{\alpha}_k \bar{\boldsymbol{A}}_k\right) \leq \bar{\sigma}\left(\boldsymbol{\alpha}_k\right)^2 \bar{\sigma}\left(\bar{\boldsymbol{A}}_k\right)^2 \bar{\sigma}\left(\boldsymbol{P}_{k+1/k}^{-1}\right), \tag{4.117}$$

and

$$\underline{\sigma}\left(\bar{\boldsymbol{A}}_k{}^T\boldsymbol{\alpha}_k\boldsymbol{C}_{k+1}^T\left(\boldsymbol{C}_{k+1}\boldsymbol{P}_{k+1/k}\boldsymbol{C}_{k+1}^T+\boldsymbol{R}_{k+1}\right)^{-1}\boldsymbol{C}_{k+1}\boldsymbol{\alpha}_k\bar{\boldsymbol{A}}_k\right)\geq$$
$$\underline{\sigma}\left(\boldsymbol{\alpha}_k\right)^2\underline{\sigma}\left(\bar{\boldsymbol{A}}_k\right)^2\underline{\sigma}\left(\boldsymbol{C}_{k+1}^-\right)^2\underline{\sigma}\left(\left(\boldsymbol{C}_{k+1}\boldsymbol{P}_{k+1/k}\boldsymbol{C}_{k+1}^T+\boldsymbol{R}_{k+1}\right)^{-1}\right)=$$
$$\frac{\underline{\sigma}\left(\boldsymbol{\alpha}_k\right)^2\underline{\sigma}\left(\bar{\boldsymbol{A}}_k\right)^2\underline{\sigma}\left(\boldsymbol{C}_{k+1}^-\right)^2}{\bar{\sigma}\left(\boldsymbol{C}_{k+1}\boldsymbol{P}_{k+1/k}\boldsymbol{C}_{k+1}^T+\boldsymbol{R}_{k+1}\right)}. \tag{4.118}$$

Applying (4.117) and (4.118) to (4.116) and then using (4.98) gives

$$\bar{\sigma}\left(\boldsymbol{\alpha}_k\right)^2\leq\underline{\sigma}\left(\boldsymbol{\alpha}_k\right)^2\frac{\underline{\sigma}\left(\bar{\boldsymbol{A}}_k\right)^2\underline{\sigma}\left(\boldsymbol{C}_{k+1}\right)^2\underline{\sigma}\left(\bar{\boldsymbol{A}}_k\boldsymbol{P}_k\bar{\boldsymbol{A}}_k{}^T+\boldsymbol{Q}_k\right)}{\bar{\sigma}\left(\boldsymbol{C}_{k+1}\boldsymbol{P}_{k+1/k}\boldsymbol{C}_{k+1}^T+\boldsymbol{R}_{k+1}\right)}+$$
$$+\frac{(1-\zeta)\underline{\sigma}\left(\bar{\boldsymbol{A}}_k\boldsymbol{P}_k\bar{\boldsymbol{A}}_k{}^T+\boldsymbol{Q}_k\right)}{\bar{\sigma}\left(\bar{\boldsymbol{A}}_k\right)^2\bar{\sigma}\left(\boldsymbol{P}_k\right)}, \tag{4.119}$$

which is equivalent to (4.108).

Thus, if the condition (4.108) is satisfied, then $\{V_k\}_{k=1}^{\infty}$ is a decreasing sequence and hence, under Assumption 4.3, the proposed observer is locally asymptotically convergent.

*Remark 4.3.* The convergence condition (4.108) is less restrictive than the solution obtained with the approach proposed in [19], which can be written as

$$\bar{\alpha}_k\leq\left(\frac{(1-\zeta)\underline{\sigma}\left(\bar{\boldsymbol{A}}_k\boldsymbol{P}_k\bar{\boldsymbol{A}}_k{}^T+\boldsymbol{Q}_k\right)}{\bar{\sigma}\left(\bar{\boldsymbol{A}}_k\right)^2\bar{\sigma}\left(\boldsymbol{P}_k\right)}\right)^{\frac{1}{2}}. \tag{4.120}$$

However, (4.108) and (4.120) become equivalent when $\boldsymbol{E}_k\neq\boldsymbol{0}$, i.e., in all cases when an unknown input is considered. This is because of the fact that the matrix $\bar{\boldsymbol{A}}_k$ is singular when $\boldsymbol{E}_k\neq\boldsymbol{0}$, which implies that $\underline{\sigma}\left(\bar{\boldsymbol{A}}_k\right)=0$. Indeed, from (4.102):

$$\begin{aligned}\bar{\boldsymbol{A}}_k&=\bar{\boldsymbol{G}}_k\boldsymbol{A}_k\\&=\left[\boldsymbol{I}-\boldsymbol{E}_k\left[(\boldsymbol{C}_{k+1}\boldsymbol{E}_k)^T\boldsymbol{C}_{k+1}\boldsymbol{E}_k\right]^{-1}(\boldsymbol{C}_{k+1}\boldsymbol{E}_k)^T\boldsymbol{C}_{k+1}\right]\boldsymbol{A}_k,\end{aligned} \tag{4.121}$$

and, under Assumption 4.4, it is evident that $\bar{\boldsymbol{A}}_k$ is singular when

$$\boldsymbol{E}_k\left[(\boldsymbol{C}_{k+1}\boldsymbol{E}_k)^T\boldsymbol{C}_{k+1}\boldsymbol{E}_k\right]^{-1}(\boldsymbol{C}_{k+1}\boldsymbol{E}_k)^T\boldsymbol{C}_{k+1} \tag{4.122}$$

is singular. The singularity of the above matrix can be easily shown with the use of (4.71), i.e.,

$$\begin{aligned}&\operatorname{rank}\left(\boldsymbol{E}_k\left[(\boldsymbol{C}_{k+1}\boldsymbol{E}_k)^T\boldsymbol{C}_{k+1}\boldsymbol{E}_k\right]^{-1}(\boldsymbol{C}_{k+1}\boldsymbol{E}_k)^T\boldsymbol{C}_{k+1}\right)\leq\\&\min\left[\operatorname{rank}(\boldsymbol{E}_k),\operatorname{rank}(\boldsymbol{C}_{k+1})\right]=q.\end{aligned} \tag{4.123}$$

*Remark 4.4.* It is clear from (4.108) that the bound of $\bar{\alpha}_k$ can be maximised by suitable settings of the instrumental matrices $\boldsymbol{Q}_k$ and $\boldsymbol{R}_k$. Indeed, $\boldsymbol{Q}_k$ should be selected in such a way as to maximise

$$\underline{\sigma}\left(\bar{\boldsymbol{A}}_k\boldsymbol{P}_k\bar{\boldsymbol{A}}_k^T+\boldsymbol{Q}_k\right). \tag{4.124}$$

To tackle this problem, let us start with a solution similar to the one proposed in [72], i.e.,

$$\boldsymbol{Q}_k=\gamma\bar{\boldsymbol{A}}_k\boldsymbol{P}_k\bar{\boldsymbol{A}}_k^T+\delta_1\boldsymbol{I}, \tag{4.125}$$

where $\gamma \geq 0$ and $\delta_1 > 0$. Substituting (4.125) into (4.124) and taking into account the fact that $\underline{\sigma}\left(\bar{\boldsymbol{A}}_k\right)=0$, it can be shown that

$$(1+\gamma)\underline{\sigma}\left(\bar{\boldsymbol{A}}_k\boldsymbol{P}_k\bar{\boldsymbol{A}}_k^T\right)+\delta_1\boldsymbol{I}=\delta_1\boldsymbol{I}. \tag{4.126}$$

Thus, this solution boils down to the classic approach with constant $\boldsymbol{Q}_k=\delta_1\boldsymbol{I}$. It is, of course, possible to set $\boldsymbol{Q}_k=\delta_1\boldsymbol{I}$ with $\delta_1$ large enough. On the other hand, it is well known that the convergence rate of such an EKF-like approach can be increased by an appropriate selection of $\boldsymbol{Q}_k$ and $\boldsymbol{R}_k$, i.e., the more accurate (near "true" values) the covariance matrices, the better the convergence rate. This means that, in the deterministic case, both of the matrices should be zero ones. Unfortunately, such an approach usually leads to the divergence of the observer as well as other computational problems. To tackle this, a compromise between the convergence and the convergence rate should be established. This can be easily done by setting $\boldsymbol{Q}_k$ as

$$\boldsymbol{Q}_k=(\gamma\boldsymbol{\varepsilon}_k^T\boldsymbol{\varepsilon}_k+\delta_1)\boldsymbol{I},\quad \boldsymbol{\varepsilon}_k=\boldsymbol{y}_k-\boldsymbol{C}_k\hat{\boldsymbol{x}}_k, \tag{4.127}$$

with $\gamma > 0$ and $\delta_1 > 0$ large and small enough, respectively. Since the form of $\boldsymbol{Q}_k$ is established, then it is possible to obtain $\boldsymbol{R}_k$ in such a way as to minimise

$$\bar{\sigma}\left(\boldsymbol{C}_{k+1}\boldsymbol{P}_{k+1/k}\boldsymbol{C}_{k+1}^T+\boldsymbol{R}_{k+1}\right). \tag{4.128}$$

To tackle this problem, let us start with the solution proposed in [19, 72]:

$$\boldsymbol{R}_{k+1}=\beta\boldsymbol{C}_{k+1}\boldsymbol{P}_{k+1/k}\boldsymbol{C}_{k+1}^T+\delta_2\boldsymbol{I}, \tag{4.129}$$

with $\beta \geq 0$ and $\delta_2 > 0$. Substituting (4.129) into (4.128) gives

$$(1+\beta)\bar{\sigma}\left(\boldsymbol{C}_{k+1}\boldsymbol{P}_{k+1/k}\boldsymbol{C}_{k+1}^T\right)+\delta_2\boldsymbol{I}. \tag{4.130}$$

Thus, from (4.130) is clear that $\boldsymbol{R}_{k+1}$ should be set as follows:

$$\boldsymbol{R}_{k+1}=\delta_2\boldsymbol{I}, \tag{4.131}$$

with $\delta_2$ small enough.

### 4.2.2 Illustrative Examples

Let us reconsider the example with an induction motor presented in Section 4.1.3. Using the same parameters and settings, the following two cases concerning the selection of $\boldsymbol{Q}_{k-1}$ and $\boldsymbol{R}_k$ were considered:

Case 1: Classic approach (constant values), i.e., $\boldsymbol{Q}_{k-1} = 0.1\boldsymbol{I}$, $\boldsymbol{R}_k = 0.1\boldsymbol{I}$
Case 2: Selection according to (4.127) and (4.131), i.e.,

$$\begin{aligned} \boldsymbol{Q}_{k-1} &= 10^{10}\boldsymbol{\varepsilon}_{k-1}^T\boldsymbol{\varepsilon}_{k-1}\boldsymbol{I} + 0.001\boldsymbol{I}, \\ \boldsymbol{R}_k &= 0.01\boldsymbol{I}, \end{aligned} \tag{4.132}$$

The results shown in Fig. 4.2 confirm the relevance of an appropriate selection of the instrumental matrices. Indeed, as can be seen, the proposed approach is superior to the classic technique of selecting the instrumental matrices $\boldsymbol{Q}_{k-1}$ and $\boldsymbol{R}_k$. Moreover, by comparing the results presented in Figs. 4.2 and 4.1 it is evident that the EUIO presented in Section 4.2 is superior to the EUIO presented in Section 4.1.

Apart from the relatively good results presented in Fig. 4.2, it can be shown that the application of stochastic robustness measures and evolutionary algorithms makes it possible to increase the convergence rate further. The details regarding such an approach as well as experiments regarding unknown input decoupling and fault diagnosis are presented in Section 6.3.

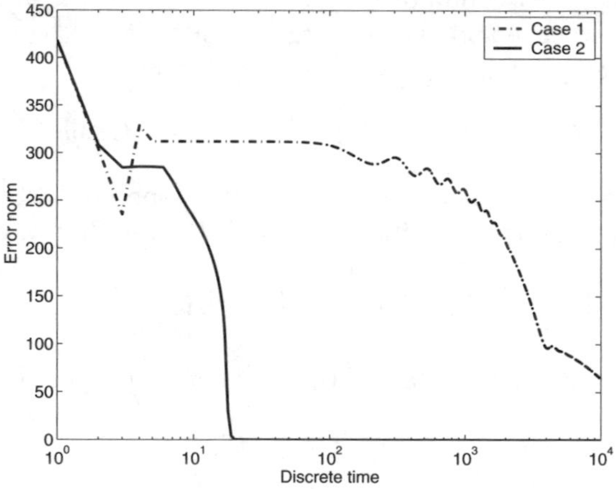


**Fig. 4.2.** State estimation error norm $\|\boldsymbol{e}_k\|_2$ for Case 1 and Case 2

## 4.3 Design of Observers and Unknown Input Observers for Lipschitz Systems

As was mentioned in Chapter 2, one way to improve the effectiveness of state estimation is to restrict the class of non-linear systems while designing observers.

Such an assumption makes it possible to avoid linearisation, which is the main tool used for the observer design purposes described in Sections 4.1 and 4.2. Section 2.2.3 presents one of such approaches, namely, observers for non-linear Lipschitz systems. Unfortunately, most of the works presented in the literature deal with continuous-time Lipschitz systems. Thus, the theory and practice concerning observers for discrete-time Lipschitz systems are significantly less mature than these for for their continuous-time counterparts. Indeed, there are few papers only [22, 172] dealing with discrete-time observers. The authors of the above works propose different parameterisations of the observer, but the common disadvantage of these approaches is that a trial-and-error procedure has to be employed that boils down to solving a large number of Lyapunov equations. Moreover, the authors do not provide convergence conditions similar to those for continuous-time observers [155, 166].

To tackle the above-mentioned difficulties, convergence criteria and the corresponding effective design procedures are presented in the subsequent part of this section. Finally, it should be pointed out that the research results portrayed in this section were originally presented in [184, 186].

### 4.3.1 Convergence Analysis

Let us consider Lipschitz systems that can be described as follows:

$$\boldsymbol{x}_{k+1} = \boldsymbol{A}\boldsymbol{x}_k + \boldsymbol{B}\boldsymbol{u}_k + \boldsymbol{h}(\boldsymbol{y}_k, \boldsymbol{u}_k) + \boldsymbol{g}\left(\boldsymbol{x}_k, \boldsymbol{u}_k\right), \tag{4.133}$$

$$\boldsymbol{y}_{k+1} = \boldsymbol{C}\boldsymbol{x}_{k+1}, \tag{4.134}$$

where $\boldsymbol{g}\left(\cdot\right)$ satisfies

$$\|\boldsymbol{g}\left(\boldsymbol{x}_1, \boldsymbol{u}\right) - \boldsymbol{g}\left(\boldsymbol{x}_2, \boldsymbol{u}\right)\|_2 \leq \gamma \|\boldsymbol{x}_1 - \boldsymbol{x}_2\|_2, \ \forall \boldsymbol{x}_1, \boldsymbol{x}_2, \boldsymbol{u}, \tag{4.135}$$

and $\gamma > 0$ stands for the Lipschitz constant.

Let us consider an observer for the system (4.133)–(4.134) described by the following equation:

$$\hat{\boldsymbol{x}}_{k+1} = \boldsymbol{A}\hat{\boldsymbol{x}}_k + \boldsymbol{B}\boldsymbol{u}_k + \boldsymbol{h}(\boldsymbol{y}_k, \boldsymbol{u}_k) + \boldsymbol{g}\left(\hat{\boldsymbol{x}}_k, \boldsymbol{u}_k\right) + \boldsymbol{K}(\boldsymbol{y}_k - \boldsymbol{C}\hat{\boldsymbol{x}}_k), \tag{4.136}$$

while $\boldsymbol{K}$ stands for the gain matrix. The subsequent part of this section shows three theorems that present three different convergence conditions of (4.136). Following Thau [166] and other researchers, let us assume that the pair $(\boldsymbol{A}, \boldsymbol{C})$ is observable. Let $\boldsymbol{P} = \boldsymbol{P}^T$, $\boldsymbol{P} > \boldsymbol{0}$ be a solution of the following Lyapunov equation:

$$\boldsymbol{Q} = \boldsymbol{P} - \boldsymbol{A}_0^T \boldsymbol{P} \boldsymbol{A}_0, \quad \boldsymbol{A}_0 = \boldsymbol{A} - \boldsymbol{K}\boldsymbol{C}, \tag{4.137}$$

with $\boldsymbol{A}_0$ being a stable matrix, i.e., $\rho(\boldsymbol{A}_0) < 1$, and $\boldsymbol{Q} = \boldsymbol{Q}^T$, $\boldsymbol{Q} > 0$.

**Theorem 4.5.** *Let us consider an observer (4.136) for the systems described by (4.133)–(4.134). If the Lipschitz constant* $\gamma$ *(cf. (4.135)) satisfies*

$$\gamma < \sqrt{\frac{\underline{\sigma}\left(\boldsymbol{Q} - \frac{1}{2}\boldsymbol{P}\right)}{\bar{\sigma}\left(\boldsymbol{P}\right)}}, \quad \boldsymbol{Q} - \frac{1}{2}\boldsymbol{P} \succ \boldsymbol{0} \tag{4.138}$$

*then the observer (4.136) is asymptotically convergent.*

*Proof.* Let us define the state estimation error for (4.136):

$$\boldsymbol{e}_k = \boldsymbol{x}_k - \hat{\boldsymbol{x}}_k, \tag{4.139}$$

and

$$\boldsymbol{s}_k = \boldsymbol{g}\left(\boldsymbol{x}_k, \boldsymbol{u}_k\right) - \boldsymbol{g}\left(\hat{\boldsymbol{x}}_k, \boldsymbol{u}_k\right). \tag{4.140}$$

Substituting (4.133)–(4.134), (4.136) and (4.140) into (4.139) gives

$$\boldsymbol{e}_{k+1} = \boldsymbol{A}_0\boldsymbol{e}_k + \boldsymbol{s}_k. \tag{4.141}$$

Let us define the following Lyapunov function:

$$V_{k+1} = \boldsymbol{e}_{k+1}^T\boldsymbol{P}\boldsymbol{e}_{k+1}, \tag{4.142}$$

and then by inserting (4.141) one can get

$$V_{k+1} = \boldsymbol{e}_k^T\boldsymbol{A}_0^T\boldsymbol{P}\boldsymbol{A}_0\boldsymbol{e}_k + 2\boldsymbol{e}_k^T\boldsymbol{A}_0^T\boldsymbol{P}\boldsymbol{s}_k + \boldsymbol{s}_k^T\boldsymbol{P}\boldsymbol{s}_k. \tag{4.143}$$

According to the Lyapunov theorem, the observer (4.136) is asymptotically convergent iff

$$\Delta V = V_{k+1} - V_k < 0. \tag{4.144}$$

Substituting (4.142) and (4.143) into (4.144) yields

$$\Delta V = \boldsymbol{e}_k^T\left[\boldsymbol{A}_0^T\boldsymbol{P}\boldsymbol{A}_0 - \boldsymbol{P}\right]\boldsymbol{e}_k + 2\boldsymbol{e}_k^T\boldsymbol{A}_0^T\boldsymbol{P}\boldsymbol{s}_k + \boldsymbol{s}_k^T\boldsymbol{P}\boldsymbol{s}_k < 0. \tag{4.145}$$

Knowing that

$$\left(\boldsymbol{P}^{\frac{1}{2}}\boldsymbol{A}_0\boldsymbol{s}_k - \boldsymbol{P}^{\frac{1}{2}}\boldsymbol{s}_k\right)^T\left(\boldsymbol{P}^{\frac{1}{2}}\boldsymbol{A}_0\boldsymbol{s}_k - \boldsymbol{P}^{\frac{1}{2}}\boldsymbol{s}_k\right) \geq 0,$$

one can obtain

$$2\boldsymbol{e}_k^T\boldsymbol{A}_0^T\boldsymbol{P}\boldsymbol{s}_k \leq \boldsymbol{e}_k^T\boldsymbol{A}_0^T\boldsymbol{P}\boldsymbol{A}_0\boldsymbol{e}_k + \boldsymbol{s}_k^T\boldsymbol{P}\boldsymbol{s}_k. \tag{4.146}$$

Inserting (4.146) into (4.145) yields

$$\Delta V \leq 2\boldsymbol{e}_k^T\left[\boldsymbol{A}_0^T\boldsymbol{P}\boldsymbol{A}_0 - \frac{1}{2}\boldsymbol{P}\right]\boldsymbol{e}_k + 2\boldsymbol{s}_k^T\boldsymbol{P}\boldsymbol{s}_k < 0. \tag{4.147}$$

Using (4.135) it can be shown that

$$\boldsymbol{s}_k^T\boldsymbol{P}\boldsymbol{s}_k \leq \gamma^2\bar{\sigma}\left(P\right)\boldsymbol{e}_k^T\boldsymbol{e}_k. \tag{4.148}$$

Substituting (4.148) into (4.147) gives

$$\Delta V \leq 2\boldsymbol{e}_k^T\left[\gamma^2\bar{\sigma}\left(P\right)\boldsymbol{I} - \left[\boldsymbol{Q} - \frac{1}{2}\boldsymbol{P}\right]\right]\boldsymbol{e}_k < 0. \tag{4.149}$$

The condition (4.149) is equivalent to

$$\gamma < \sqrt{\frac{1}{\bar{\sigma}\left(\boldsymbol{P}\right)}\frac{\boldsymbol{e}_k^T\left[\boldsymbol{Q} - \frac{1}{2}\boldsymbol{P}\right]\boldsymbol{e}_k}{\boldsymbol{e}_k^T\boldsymbol{e}_k}}. \tag{4.150}$$

Using the bound of the Rayleigh quotient, i.e., $\frac{\boldsymbol{e}_k^T\left[\boldsymbol{Q}-\frac{1}{2}\boldsymbol{P}\right]\boldsymbol{e}_k}{\boldsymbol{e}_k^T\boldsymbol{e}_k} \geq \underline{\sigma}\left(Q - \frac{1}{2}\boldsymbol{P}\right)$, it is possible to obtain (4.138), which completes the proof.

**Theorem 4.6.** *Let us consider an observer (4.136) for the systems described by (4.133)–(4.134). If the Lipschitz constant* $\gamma$ *(cf. (4.135)) satisfies*

$$\gamma < \sqrt{\frac{\underline{\sigma}\left(\boldsymbol{Q} - \boldsymbol{A}_0^T\boldsymbol{P}\boldsymbol{P}\boldsymbol{A}_0\right)}{\bar{\sigma}\left(\boldsymbol{P}\right) + 1}}, \quad \boldsymbol{Q} - \boldsymbol{A}_0^T\boldsymbol{P}\boldsymbol{P}\boldsymbol{A}_0 \succ \boldsymbol{0}, \tag{4.151}$$

*then the observer (4.136) is asymptotically convergent.*

*Proof.* Using (4.135) and the Cauchy-Schwartz inequality, it can be shown that

$$2\boldsymbol{e}_k^T\boldsymbol{A}_0^T\boldsymbol{P}\boldsymbol{s}_k \leq 2\gamma\|\boldsymbol{P}\boldsymbol{A}_0\boldsymbol{e}_k\|_2\|\boldsymbol{e}_k\|_2. \tag{4.152}$$

Applying the identity

$$\left(\|\boldsymbol{P}\boldsymbol{A}_0\boldsymbol{e}_k\|_2 - \gamma\|\boldsymbol{e}_k\|_2\right)^2 \geq 0,$$

to (4.152) yields

$$2\boldsymbol{e}_k^T\boldsymbol{A}_0^T\boldsymbol{P}\boldsymbol{s}_k \leq \boldsymbol{e}_k^T\boldsymbol{A}_0^T\boldsymbol{P}\boldsymbol{P}\boldsymbol{A}_0\boldsymbol{e}_k + \gamma^2\boldsymbol{e}_k^T\boldsymbol{e}_k. \tag{4.153}$$

Substituting (4.153) into (4.145) and then applying (4.148) leads to

$$\Delta V \leq \boldsymbol{e}_k^T\left[\gamma^2(\bar{\sigma}\left(P\right) + 1)\boldsymbol{I} - \left[\boldsymbol{Q} - \boldsymbol{A}_0^T\boldsymbol{P}\boldsymbol{P}\boldsymbol{A}_0\right]\right]\boldsymbol{e}_k < 0. \tag{4.154}$$

Finally, it is straightforward to show that (4.154) is equivalent to (4.151), which completes the proof.

**Theorem 4.7.** *Let us consider an observer (4.136) for the systems described by (4.133)–(4.134). If the Lipschitz constant* $\gamma$ *(cf. (4.135)) satisfies*

$$\gamma < \frac{\underline{\sigma}\left(\boldsymbol{Q}^{\frac{1}{2}}\right)}{\sqrt{\bar{\sigma}\left(\boldsymbol{Q}^{-\frac{1}{2}}\boldsymbol{A}_0^T\boldsymbol{P}\right)^2 + \bar{\sigma}\left(\boldsymbol{P}\right)} + \bar{\sigma}\left(\boldsymbol{Q}^{-\frac{1}{2}}\boldsymbol{A}_0^T\boldsymbol{P}\right)}, \tag{4.155}$$

*then the observer (4.136) is asymptotically convergent.*

*Proof.* Using (4.145), (4.137) and (4.148), it can be shown that the convergence condition is

$$\Delta V \leq \boldsymbol{e}_k^T\left[\gamma^2\bar{\sigma}\left(\boldsymbol{P}\right)\boldsymbol{I} - \boldsymbol{Q}\right]\boldsymbol{e}_k + 2\boldsymbol{e}_k^T\boldsymbol{A}_0^T\boldsymbol{P}\boldsymbol{s}_k < 0. \tag{4.156}$$

and hence

$$2\boldsymbol{e}_k^T\boldsymbol{A}_0^T\boldsymbol{P}\boldsymbol{s}_k < \boldsymbol{e}_k^T\left[\boldsymbol{Q} - \gamma^2\bar{\sigma}\left(\boldsymbol{P}\right)\boldsymbol{I}\right]\boldsymbol{e}_k,$$

which is equivalent to

$$2\boldsymbol{s}_k^T\boldsymbol{P}\boldsymbol{A}_0\boldsymbol{e}_k < \boldsymbol{e}_k^T\left[\boldsymbol{Q} - \gamma^2\bar{\sigma}\left(\boldsymbol{P}\right)\boldsymbol{I}\right]\boldsymbol{e}_k. \tag{4.157}$$

The inequality (4.157) can be written as follows (cf. [26]):

$$2\left(\boldsymbol{Q}^{-\frac{1}{2}}\boldsymbol{A}_0^T\boldsymbol{P}\boldsymbol{s}_k\right)^T\left(\boldsymbol{Q}^{\frac{1}{2}}\boldsymbol{e}_k\right) < \left(\boldsymbol{Q}^{\frac{1}{2}}\boldsymbol{e}_k\right)^T\left(\boldsymbol{Q}^{\frac{1}{2}}\boldsymbol{e}_k\right) - \gamma^2\bar{\sigma}\left(\boldsymbol{P}\right)\boldsymbol{e}_k^T\boldsymbol{e}_k, \tag{4.158}$$

and, hence, the convergence condition is

$$2\left\|\boldsymbol{Q}^{-\frac{1}{2}}\boldsymbol{A}_0^T\boldsymbol{P}\boldsymbol{s}_k\right\|_2 < \left\|\boldsymbol{Q}^{\frac{1}{2}}\boldsymbol{e}_k\right\|_2 - \gamma^2\bar{\sigma}\left(\boldsymbol{P}\right)\frac{\|\boldsymbol{e}_k\|_2^2}{\left\|\boldsymbol{Q}^{\frac{1}{2}}\boldsymbol{e}_k\right\|_2}. \tag{4.159}$$

Using (4.148), it can be shown that

$$\left\|\boldsymbol{Q}^{-\frac{1}{2}}\boldsymbol{A}_0^T\boldsymbol{P}\boldsymbol{s}_k\right\|_2 \le \gamma\bar{\sigma}\left(\boldsymbol{Q}^{-\frac{1}{2}}\boldsymbol{A}_0^T\boldsymbol{P}\right)\|\boldsymbol{e}_k\|_2. \tag{4.160}$$

Then, knowing that

$$\left\|\boldsymbol{Q}^{\frac{1}{2}}\boldsymbol{e}_k\right\|_2 \ge \underline{\sigma}\left(\boldsymbol{Q}^{-\frac{1}{2}}\right)\|\boldsymbol{e}_k\|_2$$

and

$$\frac{\|\boldsymbol{e}_k\|_2}{\left\|\boldsymbol{Q}^{\frac{1}{2}}\boldsymbol{e}_k\right\|_2} \le \frac{1}{\underline{\sigma}\left(\boldsymbol{Q}^{\frac{1}{2}}\right)},$$

the inequality (4.159) can be written as follows:

$$\frac{\bar{\sigma}\left(\boldsymbol{P}\right)}{\underline{\sigma}\left(\boldsymbol{Q}^{\frac{1}{2}}\right)}\gamma^2 + 2\gamma\bar{\sigma}\left(\boldsymbol{Q}^{-\frac{1}{2}}\boldsymbol{A}_0^T\boldsymbol{P}\right) - \underline{\sigma}\left(\boldsymbol{Q}^{\frac{1}{2}}\right) < 0. \tag{4.161}$$

Since (4.161) contains a quadratic function, then it is clear that

$$\gamma < \left(\sqrt{\bar{\sigma}\left(\boldsymbol{Q}^{-\frac{1}{2}}\boldsymbol{A}_0^T\boldsymbol{P}\right)^2 + \bar{\sigma}\left(\boldsymbol{P}\right)} - \bar{\sigma}\left(\boldsymbol{Q}^{-\frac{1}{2}}\boldsymbol{A}_0^T\boldsymbol{P}\right)\right)\frac{\underline{\sigma}\left(\boldsymbol{Q}^{\frac{1}{2}}\right)}{\bar{\sigma}\left(\boldsymbol{P}\right)}. \tag{4.162}$$

Finally, using the identity

$$\begin{aligned}&\left(\sqrt{\bar{\sigma}\left(\boldsymbol{Q}^{-\frac{1}{2}}\boldsymbol{A}_0^T\boldsymbol{P}\right)^2 + \bar{\sigma}\left(\boldsymbol{P}\right)} - \bar{\sigma}\left(\boldsymbol{Q}^{-\frac{1}{2}}\boldsymbol{A}_0^T\boldsymbol{P}\right)\right)\cdot\\ &\cdot\left(\sqrt{\bar{\sigma}\left(\boldsymbol{Q}^{-\frac{1}{2}}\boldsymbol{A}_0^T\boldsymbol{P}\right)^2 + \bar{\sigma}\left(\boldsymbol{P}\right)} + \bar{\sigma}\left(\boldsymbol{Q}^{-\frac{1}{2}}\boldsymbol{A}_0^T\boldsymbol{P}\right)\right)\\ &= \bar{\sigma}\left(\boldsymbol{P}\right),\end{aligned}$$

the inequality (4.162) can be transformed into (4.155), which completes the proof.

*Remark 4.8.* The convergence criteria described by the above theorems are obtained by eliminating the term

$$2\boldsymbol{e}_k^T\boldsymbol{A}_0^T\boldsymbol{P}\boldsymbol{s}_k$$

from (4.145) in three distinct ways. This means that the obtained criteria are relatively conservative and the scale of this conservatism is strongly related to the inaccuracy of a given elimination technique.

*Remark 4.9.* There is no doubt that there are particular choices of $\boldsymbol{Q}$ which will bring forth the least conservative bounds (4.138), (4.151) and (4.155). Unfortunately, the structural relation between $\boldsymbol{P}$ and $\boldsymbol{Q}$ of (4.137) cannot be resolved without first solving the Lyapunov equation. This is the main reason why it is impossible to chose one criterion that gives the least conservative bound of $\gamma$ for an arbitrary matrix $\boldsymbol{Q}$.

*Remark 4.10.* Unfortunately, (4.138), (4.151) and (4.155) may merely serve as methods of checking the convergence, but the gain matrix $\boldsymbol{K}$ has to be determined beforehand. This means that the design procedure boils down to selecting various gain matrices $\boldsymbol{K}$, solving the Lyapunov equation (4.137), and then checking the convergence conditions (4.138), (4.151) and (4.155). There is no doubt that this is an ineffective and inconvenient approach.

Taking into account the above remarks, the objective of the subsequent section is to develop three different design procedures that are based on (4.138), (4.151) and (4.155).

### 4.3.2 Design Procedures

#### Design procedure I

It can easily be shown that (4.149) is equivalent to

$$\gamma^2 \bar{\sigma}\left(\boldsymbol{P}\right)\boldsymbol{I} + \boldsymbol{A}_0^T \boldsymbol{P} \boldsymbol{A}_0 - \frac{1}{2}\boldsymbol{P} \prec \boldsymbol{0}. \tag{4.163}$$

Assuming that $\bar{\sigma}\left(\boldsymbol{P}\right) < \beta$, $\beta > 0$, and knowing that $\bar{\sigma}\left(\boldsymbol{P}\right) < \beta$ is equivalent to $\beta - \beta^{-1}\boldsymbol{P}\boldsymbol{P} \succ \boldsymbol{0}$, which can be written in the following LMI form:

$$\begin{bmatrix} \beta\boldsymbol{I} & \boldsymbol{P} \\ \boldsymbol{P} & \beta\boldsymbol{I} \end{bmatrix} \succ \boldsymbol{0}, \quad \beta > 0, \quad \boldsymbol{P} \succ \boldsymbol{0}, \tag{4.164}$$

(4.163) can be transformed into a set of inequalities:

$$\gamma^2 \beta \boldsymbol{I} + \boldsymbol{A}_0^T \boldsymbol{P} \boldsymbol{A}_0 - \frac{1}{2}\boldsymbol{P} \prec \boldsymbol{0}, \tag{4.165}$$

and (4.164). The inequality (4.165) can be written in the following form:

$$\begin{bmatrix} \frac{1}{2}\boldsymbol{P} - \gamma^2\beta\boldsymbol{I} & \boldsymbol{A}_0^T \\ \boldsymbol{A}_0 & \boldsymbol{P}^{-1} \end{bmatrix} \succ \boldsymbol{0}, \tag{4.166}$$

which is equivalent to

$$\begin{bmatrix} \boldsymbol{I} & \boldsymbol{0} \\ \boldsymbol{0} & \boldsymbol{P} \end{bmatrix} \begin{bmatrix} \frac{1}{2}\boldsymbol{P} - \gamma^2\beta\boldsymbol{I} & \boldsymbol{A}_0^T \\ \boldsymbol{A}_0 & \boldsymbol{P}^{-1} \end{bmatrix} \begin{bmatrix} \boldsymbol{I} & \boldsymbol{0} \\ \boldsymbol{0} & \boldsymbol{P} \end{bmatrix} \succ \boldsymbol{0}. \tag{4.167}$$

Finally, (4.167) can be written in the following form:

$$\begin{bmatrix} \frac{1}{2}\boldsymbol{P} - \gamma^2\beta\boldsymbol{I} & \boldsymbol{A}_0^T\boldsymbol{P} \\ \boldsymbol{P}\boldsymbol{A}_0 & \boldsymbol{P} \end{bmatrix} \succ \boldsymbol{0}. \tag{4.168}$$

Substituting $\boldsymbol{K} = \boldsymbol{P}^{-1}\boldsymbol{L}$ into (4.168) yields the following LMI:

$$\begin{bmatrix} \frac{1}{2}\boldsymbol{P} - \gamma^2\beta\boldsymbol{I} & \boldsymbol{A}^T\boldsymbol{P} - \boldsymbol{C}^T\boldsymbol{L}^T \\ \boldsymbol{P}\boldsymbol{A} - \boldsymbol{L}\boldsymbol{C} & \boldsymbol{P} \end{bmatrix} \succ \mathbf{0}. \tag{4.169}$$

Thus, the design procedure can be summarised as follows:

Step 1: Obtain $\gamma$ for (4.133)–(4.134).
Step 2: Solve a set of LMIs: (4.164) and (4.169).
Step 3: Obtain the gain matrix $\boldsymbol{K} = \boldsymbol{P}^{-1}\boldsymbol{L}$.

In spite of the simplicity and effectiveness of the proposed approach, it cannot be directly applied to determine $\boldsymbol{K}$ maximising $\gamma$ for which the observer (4.136) is convergent. The objective of the subsequent part of this section is to tackle the above-defined task. It can be observed that (4.169) can be transformed into the following form:

$$\begin{bmatrix} -\frac{1}{2}\boldsymbol{P} & \boldsymbol{C}^T\boldsymbol{L}^T - \boldsymbol{A}^T\boldsymbol{P} \\ \boldsymbol{L}\boldsymbol{C} - \boldsymbol{P}\boldsymbol{A} & -\boldsymbol{P} \end{bmatrix} \prec \lambda \begin{bmatrix} \beta\boldsymbol{I} & \mathbf{0} \\ \mathbf{0} & \mathbf{0} \end{bmatrix} \tag{4.170}$$

where $\lambda = -\gamma^2$. Thus, the task can be reduced to a generalised eigenvalue minimisation problem [20, 62] that can be formulated as follows:

$$\min_{\boldsymbol{P},\boldsymbol{L},\beta} \lambda,$$

under the LMI constraints (4.164) and (4.170). As can be observed, the right hand side of (4.170) is semi-positive definite. The positivity of the right hand side of (4.170) is usually required for the well-posedness of the task and the applicability of the polynomial-time interior point methods [62]. For a simple remedy to this problem, the reader is referred to [62, p. 8-41] (see also Appendix).

It should be also strongly underlined that when the optimisation problem described by *Steps I–III* (or in the form of the generalised eigenvalue minimisation problem) cannot be solved due to its infeasibility, then the only way out is to transform the original description of the system into an equivalent one with a smaller Lipschitz constant. Some guidance regarding such a strategy is given in [1, 150]. Thus, due the the observability assumption, the algorithm is guaranteed to converge as $\gamma \to 0$. This is, of course, a common drawback of the existing approaches to the design of observers for non-linear Lipschitz systems (cf. [1, 142, 150]).

### Design procedure II

It can easily be shown that (4.154) is equivalent to

$$\gamma^2(\bar{\sigma}(\boldsymbol{P}) + 1)\boldsymbol{I} + \boldsymbol{A}_0^T\boldsymbol{P}\boldsymbol{A}_0 + \boldsymbol{A}_0^T\boldsymbol{P}\boldsymbol{P}\boldsymbol{A}_0 - \boldsymbol{P} \prec \mathbf{0}, \tag{4.171}$$

Assuming that $\bar{\sigma}(\boldsymbol{P}) < \beta$, $\beta > 0$ and $\boldsymbol{A}_0^T\boldsymbol{P}\boldsymbol{P}\boldsymbol{A}_0 \prec \boldsymbol{X}$, $\boldsymbol{X} = \boldsymbol{X}^T$, which can be expressed as

$$\begin{bmatrix} \boldsymbol{X} & \boldsymbol{A}_0^T\boldsymbol{P} \\ \boldsymbol{P}\boldsymbol{A}_0 & \boldsymbol{I} \end{bmatrix} \succ \mathbf{0}, \tag{4.172}$$

the inequality (4.171) can be written as follows:

$$\begin{bmatrix} \boldsymbol{P} - \gamma^2(\beta+1)\boldsymbol{I} - \boldsymbol{X} & \boldsymbol{A}_0^T\boldsymbol{P} \\ \boldsymbol{P}\boldsymbol{A}_0 & \boldsymbol{P} \end{bmatrix} \succ \boldsymbol{0}. \tag{4.173}$$

Substituting $\boldsymbol{K} = \boldsymbol{P}^{-1}\boldsymbol{L}$ into (4.172) and (4.173) yields the following set of LMIs:

$$\begin{bmatrix} \boldsymbol{X} & \boldsymbol{A}^T\boldsymbol{P} - \boldsymbol{C}^T\boldsymbol{L}^T \\ \boldsymbol{P}\boldsymbol{A} - \boldsymbol{L}\boldsymbol{C} & \boldsymbol{I} \end{bmatrix} \succ \boldsymbol{0}, \tag{4.174}$$

and

$$\begin{bmatrix} \boldsymbol{P} - \gamma^2(\beta+1)\boldsymbol{I} - \boldsymbol{X} & \boldsymbol{A}^T\boldsymbol{P} - \boldsymbol{C}^T\boldsymbol{L}^T \\ \boldsymbol{P}\boldsymbol{A} - \boldsymbol{L}\boldsymbol{C} & \boldsymbol{P} \end{bmatrix} \succ \boldsymbol{0}. \tag{4.175}$$

Thus, the new design procedure can be summarised as follows:

Step 1: Obtain $\gamma$ for (4.133)–(4.134).
Step 2: Solve a set of LMIs: (4.164), (4.174), and (4.175).
Step 3: Obtain the gain matrix $\boldsymbol{K} = \boldsymbol{P}^{-1}\boldsymbol{L}$.

Similarly as for the design procedure I, the selection of $\boldsymbol{K}$ maximising $\gamma$ for which the observer (4.136) is convergent can be formulated as the generalised eigenvalue minimisation problem:

$$\min_{\boldsymbol{P},\boldsymbol{L},\boldsymbol{X},\beta} \lambda,$$

under the LMI constraints (4.164), (4.174), and

$$\begin{bmatrix} \boldsymbol{X} - \boldsymbol{P} & \boldsymbol{C}^T\boldsymbol{L}^T - \boldsymbol{A}^T\boldsymbol{P} \\ \boldsymbol{L}\boldsymbol{C} - \boldsymbol{P}\boldsymbol{A} & -\boldsymbol{P} \end{bmatrix} \prec \lambda \begin{bmatrix} (\beta+1)\boldsymbol{I} & \boldsymbol{0} \\ \boldsymbol{0} & \boldsymbol{0} \end{bmatrix}, \tag{4.176}$$

where (4.176) is obtained by suitably rearranging (4.175), and $\lambda = -\gamma^2$.

### Design procedure III

The inequality (4.161) can be transformed into an equivalent form:

$$\bar{\sigma}(\boldsymbol{P})\gamma^2 + 2\gamma\underline{\sigma}\left(\boldsymbol{Q}^{\frac{1}{2}}\right)\bar{\sigma}\left(\boldsymbol{Q}^{-\frac{1}{2}}\boldsymbol{A}_0^T\boldsymbol{P}\right) - \underline{\sigma}(\boldsymbol{Q}) < 0. \tag{4.177}$$

Knowing that

$$\underline{\sigma}\left(\boldsymbol{Q}^{\frac{1}{2}}\right)\bar{\sigma}\left(\boldsymbol{Q}^{-\frac{1}{2}}\boldsymbol{A}_0^T\boldsymbol{P}\right) \le \bar{\sigma}\left(\boldsymbol{A}_0^T\boldsymbol{P}\right), \tag{4.178}$$

the inequality (4.177) can be written as:

$$\bar{\sigma}(\boldsymbol{P})\gamma^2 + 2\gamma\bar{\sigma}\left(\boldsymbol{A}_0^T\boldsymbol{P}\right) - \underline{\sigma}(\boldsymbol{Q}) < 0. \tag{4.179}$$

Assuming that $\bar{\sigma}(\boldsymbol{P}) < \beta$, $\beta > 0$ and $\bar{\sigma}\left(\boldsymbol{A}_0^T\boldsymbol{P}\right) < \delta$, $\delta > 0$, which can be expressed as

$$\begin{bmatrix} \delta & \boldsymbol{A}_0^T\boldsymbol{P} \\ \boldsymbol{P}\boldsymbol{A}_0 & \delta \end{bmatrix} \succ \boldsymbol{0}, \quad \delta > 0, \tag{4.180}$$

now it is straightforward to show that (4.179) can be represented by

$$\boldsymbol{P} - \boldsymbol{A}_0^T \boldsymbol{P} \boldsymbol{A}_0 - \gamma^2 \beta \boldsymbol{I} - 2\gamma\delta \boldsymbol{I} \succ \boldsymbol{0}. \tag{4.181}$$

Thus, the inequality (4.181) can be written as follows:

$$\begin{bmatrix} \boldsymbol{P} - \gamma^2\beta\boldsymbol{I} - 2\gamma\delta\boldsymbol{I} & \boldsymbol{A}_0^T\boldsymbol{P} \\ \boldsymbol{P}\boldsymbol{A}_0 & \boldsymbol{P} \end{bmatrix} \succ \boldsymbol{0}. \tag{4.182}$$

Substituting $\boldsymbol{K} = \boldsymbol{P}^{-1}\boldsymbol{L}$ into (4.180) and (4.182) yields the following set of LMIs:

$$\begin{bmatrix} \delta & \boldsymbol{A}^T\boldsymbol{P} - \boldsymbol{C}^T\boldsymbol{L}^T \\ \boldsymbol{P}\boldsymbol{A} - \boldsymbol{L}\boldsymbol{C} & \delta \end{bmatrix} \succ \boldsymbol{0}, \quad \delta > 0, \tag{4.183}$$

and

$$\begin{bmatrix} \boldsymbol{P} - \gamma^2\beta\boldsymbol{I} - 2\gamma\delta\boldsymbol{I} & \boldsymbol{A}^T\boldsymbol{P} - \boldsymbol{C}^T\boldsymbol{L}^T \\ \boldsymbol{P}\boldsymbol{A} - \boldsymbol{L}\boldsymbol{C} & \boldsymbol{P} \end{bmatrix} \succ \boldsymbol{0}. \tag{4.184}$$

Thus, the third design procedure can be summarised as follows:

Step 1: Obtain $\gamma$ for (4.133)–(4.134).
Step 2: Solve a set of LMIs: (4.164), (4.183), and (4.184).
Step 3: Obtain the gain matrix $\boldsymbol{K} = \boldsymbol{P}^{-1}\boldsymbol{L}$.

Similarly as for the first and second design procedures, the selection of $\boldsymbol{K}$ maximising $\gamma$ for which the observer (4.136) is convergent can be formulated as the generalised eigenvalue minimisation problem. First, let us assume that

$$-\boldsymbol{X} \prec \lambda\beta\boldsymbol{I}, \quad \boldsymbol{X} \succ \boldsymbol{0}, \tag{4.185}$$

where $\boldsymbol{X} = \boldsymbol{X}^T$, $\lambda = -\gamma$. Thus, the inequality (4.184) can be expressed as

$$\begin{bmatrix} -\boldsymbol{P} & \boldsymbol{C}^T\boldsymbol{L}^T - \boldsymbol{A}^T\boldsymbol{P} \\ \boldsymbol{L}\boldsymbol{C} - \boldsymbol{P}\boldsymbol{A} & -\boldsymbol{P} \end{bmatrix} \prec \lambda \begin{bmatrix} \boldsymbol{X} + 2\delta\boldsymbol{I} & \boldsymbol{0} \\ \boldsymbol{0} & \boldsymbol{0} \end{bmatrix}. \tag{4.186}$$

Finally, the generalised eigenvalue minimisation problem boils down to

$$\min_{\boldsymbol{P},\boldsymbol{L},\boldsymbol{X},\beta,\delta} \lambda,$$

under the LMI constraints (4.164), (4.183), (4.185)–(4.186).

### 4.3.3 Design of an Unknown Input Observer

The purpose of the subsequent part of this section is to present a straightforward approach for extending the techniques proposed in the preceding sections to discrete-time Lipschitz systems with unknown inputs, which can be described as follows:

$$\boldsymbol{x}_{k+1} = \boldsymbol{A}\boldsymbol{x}_k + \boldsymbol{B}\boldsymbol{u}_k + \boldsymbol{h}(\boldsymbol{y}_k, \boldsymbol{u}_k) + \boldsymbol{g}(\boldsymbol{x}_k, \boldsymbol{u}_k) + \boldsymbol{E}\boldsymbol{d}_k, \tag{4.187}$$

$$\boldsymbol{y}_{k+1} = \boldsymbol{C}\boldsymbol{x}_{k+1}. \tag{4.188}$$

In order to use the techniques described in the preceding sections for state estimation of the system (4.187)–(4.188), it is necessary to introduce some modifications concerning the unknown input.

The derivation presented in the subsequent part of this section is based on the selected results presented in Section 4.2. Similarly as in Sections 4.1 and 4.2, let us assume that

$$\mathrm{rank}(\boldsymbol{C}\boldsymbol{E}) = \mathrm{rank}(\boldsymbol{E}) = q, \tag{4.189}$$

(see [27, p. 72, Lemma 3.1] for a comprehensive explanation). If the condition (4.189) is satisfied, then it is possible to calculate $\boldsymbol{H} = (\boldsymbol{C}\boldsymbol{E})^{+} = \left[(\boldsymbol{C}\boldsymbol{E})^{T}\boldsymbol{C}\boldsymbol{E}\right]^{-1}(\boldsymbol{C}\boldsymbol{E})^{T}$, where $(\cdot)^{+}$ stands for the pseudo-inverse of its argument. Thus, let us use the first of the above mentioned techniques for designing UIOs [91]. By multiplying (4.188) by $\boldsymbol{H}$ and then inserting (4.187), it is straightforward to show that

$$\boldsymbol{d}_k = \boldsymbol{H}\left[\boldsymbol{y}_{k+1} - \boldsymbol{C}\left[\boldsymbol{A}\boldsymbol{x}_k + \boldsymbol{B}\boldsymbol{u}_k + \boldsymbol{h}(\boldsymbol{y}_k, \boldsymbol{u}_k) + \boldsymbol{g}\left(\boldsymbol{x}_k, \boldsymbol{u}_k\right)\right]\right]. \tag{4.190}$$

Substituting (4.190) into (4.187) gives

$$\boldsymbol{x}_{k+1} = \bar{\boldsymbol{A}}\boldsymbol{x}_k + \bar{\boldsymbol{B}}\boldsymbol{u}_k + \bar{\boldsymbol{h}}\left(\boldsymbol{u}_k, \boldsymbol{y}_k\right) + \bar{\boldsymbol{g}}\left(\boldsymbol{x}_k, \boldsymbol{u}_k\right) + \bar{\boldsymbol{E}}\boldsymbol{y}_{k+1}, \tag{4.191}$$

where

$$\begin{aligned}
&\bar{\boldsymbol{A}} = \bar{\boldsymbol{G}}\boldsymbol{A},\ \bar{\boldsymbol{B}} = \bar{\boldsymbol{G}}\boldsymbol{B},\ \bar{\boldsymbol{g}}\left(\cdot\right) = \bar{\boldsymbol{G}}\boldsymbol{g}\left(\cdot\right),\ \bar{\boldsymbol{h}}\left(\cdot\right) = \bar{\boldsymbol{G}}\boldsymbol{h}(\cdot)\\
&\bar{\boldsymbol{G}} = \boldsymbol{I} - \boldsymbol{E}\boldsymbol{H}\boldsymbol{C},\ \bar{\boldsymbol{E}} = \boldsymbol{E}\boldsymbol{H}.
\end{aligned}$$

Thus, the unknown input observer for (4.187)–(4.188) is given as follows:

$$\begin{aligned}
\hat{\boldsymbol{x}}_{k+1} =& \bar{\boldsymbol{A}}\hat{\boldsymbol{x}}_k + \bar{\boldsymbol{B}}\boldsymbol{u}_k + \bar{\boldsymbol{h}}\left(\boldsymbol{u}_k, \boldsymbol{y}_k\right) + \bar{\boldsymbol{g}}\left(\boldsymbol{x}_k, \boldsymbol{u}_k\right) +\\
&+ \bar{\boldsymbol{E}}\boldsymbol{y}_{k+1} + \boldsymbol{K}(\boldsymbol{y}_k - \boldsymbol{C}\hat{\boldsymbol{x}}_k).
\end{aligned} \tag{4.192}$$

Now, let us consider the first of the above-mentioned approaches, which can be used for designing the UIO [27]. For the sake of notational simplicity, let us start with the UIO for linear discrete-time systems:

$$\begin{aligned}
\boldsymbol{x}_{k+1} &= \boldsymbol{A}\boldsymbol{x}_k + \boldsymbol{B}\boldsymbol{u}_k + \boldsymbol{E}\boldsymbol{d}_k,\\
\boldsymbol{y}_{k+1} &= \boldsymbol{C}\boldsymbol{x}_{k+1},
\end{aligned} \tag{4.193}$$

which can be described as follows:

$$\boldsymbol{s}_{k+1} = \boldsymbol{F}\boldsymbol{s}_k + \boldsymbol{T}\boldsymbol{B}\boldsymbol{u}_k + \boldsymbol{K}_1\boldsymbol{y}_k, \tag{4.194}$$

$$\hat{\boldsymbol{x}}_{k+1} = \boldsymbol{s}_{k+1} + \boldsymbol{H}_1\boldsymbol{y}_{k+1}, \tag{4.195}$$

with

$$\boldsymbol{K}_1 = \boldsymbol{K} + \boldsymbol{K}_2, \tag{4.196}$$

$$\boldsymbol{E} = \boldsymbol{H}_1\boldsymbol{C}\boldsymbol{E}, \tag{4.197}$$

$$\boldsymbol{T} = \boldsymbol{I} - \boldsymbol{H}_1\boldsymbol{C}, \tag{4.198}$$

$$\boldsymbol{F} = \boldsymbol{T}\boldsymbol{A} - \boldsymbol{K}\boldsymbol{C}, \tag{4.199}$$

$$\boldsymbol{K}_2 = \boldsymbol{F}\boldsymbol{H}_1. \tag{4.200}$$

By substituting (4.195) into (4.194) and then using (4.198), (4.199) and (4.200), it can be shown that:

$$\begin{aligned}\hat{\boldsymbol{x}}_{k+1} =&\boldsymbol{A}\hat{\boldsymbol{x}}_k + \boldsymbol{B}\boldsymbol{u}_k - \boldsymbol{H}_1\boldsymbol{C}[\boldsymbol{A}\hat{\boldsymbol{x}}_k + \boldsymbol{B}\boldsymbol{u}_k] - \boldsymbol{K}\boldsymbol{C}\hat{\boldsymbol{x}}_k - \boldsymbol{F}\boldsymbol{H}_1\boldsymbol{y}_k + \\ &+ [\boldsymbol{K} + \boldsymbol{F}\boldsymbol{H}_1]\boldsymbol{y}_k + \boldsymbol{H}_1\boldsymbol{y}_{k+1},\end{aligned} \tag{4.201}$$

or, equivalently,

$$\hat{\boldsymbol{x}}_{k+1} = \hat{\boldsymbol{x}}_{k+1/k} + \boldsymbol{H}_1(\boldsymbol{y}_{k+1} - \boldsymbol{C}\hat{\boldsymbol{x}}_{k+1/k}) + \boldsymbol{K}(\boldsymbol{y}_k - \boldsymbol{C}\hat{\boldsymbol{x}}_k), \tag{4.202}$$

where

$$\hat{\boldsymbol{x}}_{k+1/k} = \boldsymbol{A}\hat{\boldsymbol{x}}_k + \boldsymbol{B}\boldsymbol{u}_k. \tag{4.203}$$

Substituting the solution of (4.197), i.e., $\boldsymbol{H}_1 = \boldsymbol{E}\boldsymbol{H}$ into (4.202) yields:

$$\hat{\boldsymbol{x}}_{k+1} = [\boldsymbol{I} - \boldsymbol{E}\boldsymbol{H}\boldsymbol{C}]\hat{\boldsymbol{x}}_{k+1/k} + \boldsymbol{E}\boldsymbol{H}\boldsymbol{y}_{k+1} + \boldsymbol{K}(\boldsymbol{y}_k - \boldsymbol{C}\hat{\boldsymbol{x}}_k). \tag{4.204}$$

Thus, in order to use (4.204) for (4.187)–(4.188) it is necessary to replace (4.203) by

$$\hat{\boldsymbol{x}}_{k+1/k} = \boldsymbol{A}\hat{\boldsymbol{x}}_k + \boldsymbol{B}\boldsymbol{u}_k + \boldsymbol{h}(\boldsymbol{y}_k, \boldsymbol{u}_k) + \boldsymbol{g}\left(\hat{\boldsymbol{x}}_k, \boldsymbol{u}_k\right). \tag{4.205}$$

Similarly as was the case in Section 4.2, by substituting (4.205) into (4.204) and then comparing it with (4.192) it can be seen that the observer structures being considered are identical. On the other hand, it should be clearly pointed out that they were designed in a significantly different way.

Since the observer structure is established, then it is possible to describe its design procedure.

A simple comparison of (4.133) and (4.191) leads to the conclusion that the observer (4.192) can be designed with one of the techniques proposed in Section 4.3.2, taking into account the fact that (cf. (4.135)):

$$\|\bar{\boldsymbol{g}}\left(\boldsymbol{x}_1, \boldsymbol{u}\right) - \bar{\boldsymbol{g}}\left(\boldsymbol{x}_2, \boldsymbol{u}\right)\|_2 \leq \bar{\gamma}\|\boldsymbol{x}_1 - \boldsymbol{x}_2\|_2, \quad \forall \boldsymbol{x}_1, \boldsymbol{x}_2, \boldsymbol{u}, \tag{4.206}$$

and assuming that the pair $(\bar{\boldsymbol{A}}, \boldsymbol{C})$ is observable.

### 4.3.4 Experimental Results

#### Observer design and state estimation

The main objective of the present section is to compare the performance of the three different design procedures (proposed in Section 4.3.2).

First, the problem is to obtain the gain matrix $\boldsymbol{K}$ maximising $\gamma$ (for which the observer (4.136) is convergent) for the systems given by

$$\boldsymbol{A} = \begin{bmatrix} 0.2 & 0.01 \\ 0.1 & 0.2 \end{bmatrix}, \quad \boldsymbol{C} = [1\, 0], \tag{4.207}$$

and

$$\boldsymbol{A} = \begin{bmatrix} 0.137 & 0.199 & 0.284 \\ 0.0118 & 0.299 & 0.47 \\ 0.894 & 0.661 & 0.065 \end{bmatrix}, \quad \boldsymbol{C} = \begin{bmatrix} 1 & 0 & 0 \\ 0 & 1 & 0 \end{bmatrix}. \tag{4.208}$$

To tackle this problem the approaches presented in Section 4.3.2 were implemented with MATLAB®. One of the functions that implements the design procedure II is presented and carefully discussed in Appendix.

**Table 4.1.** Maximum $\gamma$ for (4.207) and (4.208)

| Design procedure | $\gamma$ for (4.207) | $\gamma$ for (4.208) |
|---|---|---|
| I | 0.6765 | 0.5563 |
| II | 0.7998 | 0.6429 |
| III | 0.7916 | 0.5422 |

The obtained results are presented in Tab. 4.1. It can be observed that the maximum difference between the maximum Lipschitz constant obtained with the proposed design procedures is greater than 15%. Apart from the fact that the second design procedure gave the best results, it is probably impossible to prove that this is the best choice for all systems. The above results confirm Remark 4.9, i.e., it is very hard to chose a priori a criterion that gives the least conservative bound of $\gamma$. It is also worth noting that, contrary to the approaches presented in the literature (see, e.g., [1, 149, 150, 172]), the proposed procedures provide the gain matrix $\boldsymbol{K}$ that is a global solution of the $\gamma$ maximisation problem.

Now let us consider a one-link manipulator with revolute joints actuated by a DC motor [150] described by the continuous counterpart of (4.133)–(4.134) with the following parameters:

$$\boldsymbol{A} = \begin{bmatrix} 0 & 1 & 0 & 0 \\ -48.6 & -1.25 & 48.6 & 0 \\ 0 & 0 & 0 & 10 \\ 1.95 & 0 & -1.95 & 0 \end{bmatrix},$$

$$\boldsymbol{B} = [0\,21.6\,0\,0]^T, \quad \boldsymbol{C} = \begin{bmatrix} 1 & 0 & 0 & 0 \\ 0 & 1 & 0 & 0 \end{bmatrix},$$

$$\boldsymbol{g}(\boldsymbol{x}(t), \boldsymbol{u}(t)) = [0\,0\,0\; - 0.333\sin(x_3)], \quad \boldsymbol{h}(\boldsymbol{y}(t), \boldsymbol{u}(t)) = \boldsymbol{0}, \tag{4.209}$$

where $x_1(t)$ stands for the angular rotation of the motor, $x_2(t)$ is the angular velocity of the motor, $x_3(t)$ is the angular position of the link, and $x_4(t)$ is the angular velocity of the link.

The discrete-time counterpart (4.133)–(4.134) of (4.209) was obtained by using the Euler discretisation of a step size $\tau = 0.01$. The input signal was given by $u_k = \sin(2\pi\tau k)$, while the initial condition for the observer and the system were $\hat{\boldsymbol{x}}_0 = \boldsymbol{1}$ and $\boldsymbol{x}_0 = \boldsymbol{0}$, respectively.

The first objective was to compare the performance of the three different design procedures (proposed in Section 4.3.2). In particular, the problem was to obtain the gain matrix $\boldsymbol{K}$ maximising $\gamma$.

As can be easily observed, the Lipschitz constant $\gamma = \tau 0.333$. The following maximum values of $\gamma$ were obtained for the consecutive design procedures (i.e., I, II, III): $\gamma = 0.0329$, $\gamma = 0.0802$, $\gamma = 0.0392$. This means that acceptable $\gamma$ (provided by the second design procedure) is more than 24 times larger than actual $\gamma = 0.00333$. Similarly as in the preceding examples, the best results were achieved for the second design procedure. The resulting gain matrix $\boldsymbol{K}$ is

$$\boldsymbol{K} = \begin{bmatrix} 1.0000 & 0.0100 \\ -0.4860 & 1.7926 \\ 0 & 1.9822 \\ 0.0195 & 3.2371 \end{bmatrix}. \tag{4.210}$$

For the purpose of comparison, a continuous-time observer

$$\dot{\hat{\boldsymbol{x}}}(t) = \boldsymbol{A}\hat{\boldsymbol{x}}(t) + \boldsymbol{B}\boldsymbol{u}(t) + \boldsymbol{g}\left(\hat{\boldsymbol{x}}(t), \boldsymbol{u}(t)\right) + \boldsymbol{K}(\boldsymbol{y}(t) - \boldsymbol{C}\hat{\boldsymbol{x}}(t)) \tag{4.211}$$

designed by Rajamani and Cho [150] was employed. They obtained the following gain matrix $\boldsymbol{K}$ for (4.209):

$$\boldsymbol{K} = \begin{bmatrix} 0.8307 & 0.4514 \\ 0.4514 & 6.2310 \\ 0.8238 & 1.3072 \\ 0.0706 & 0.2574 \end{bmatrix}, \tag{4.212}$$

which is also utilised in this section.

Figures 4.3–4.6 show the results of state estimation. As can be observed, the state estimates obtained with the proposed observer converge rapidly to the corresponding true values (compare especially the estimates for $k = 0, \ldots, 40$ exposed by the plots on the right in Figs. 4.3–4.6). Indeed, it can be easily seen that the proposed observer is superior to the one designed with the design procedure proposed in [150]. This superiority can be clearly seen in Fig. 4.7, which exposes the evolution of the norm of the state estimation error.

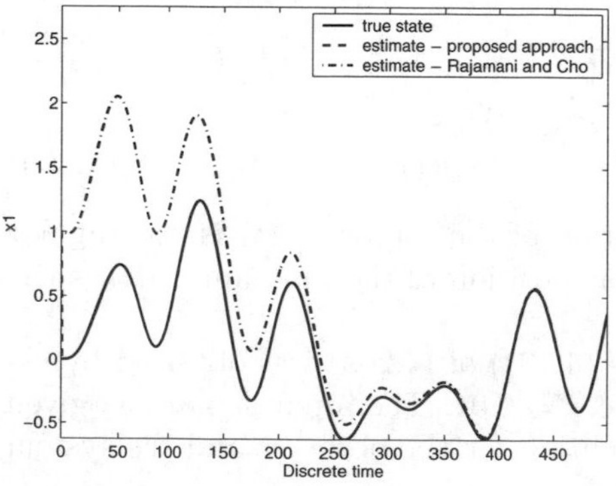


(a)

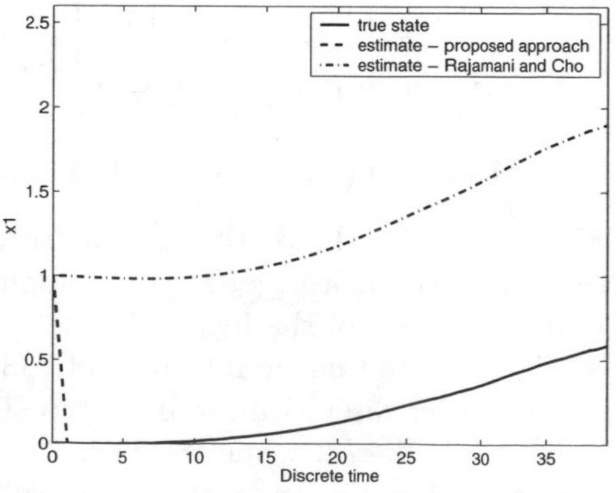


(b)

**Fig. 4.3.** Angular rotation of the motor $x_1$ and its estimates for the entire simulation (a) and first 41 samples (b)

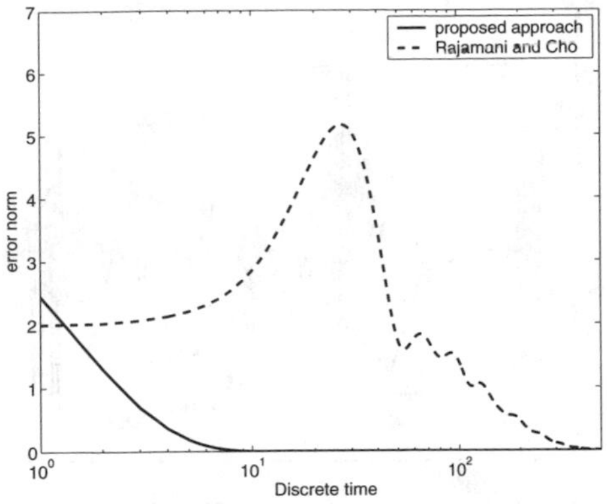


**Fig. 4.7.** Norm of the state estimation error $\|e_k\|_2$

impossible to detect the actuator fault with the conventional observer and the fixed threshold (presented in the figure). Contrarily, it is straightforward to assign a fixed threshold for the residual generated with the UIO, and then to detect the actuator fault with $z_{1,k} = y_{1,k} - \hat{y}_{1,k}$ (Fig. 4.8).

## 4.4 Concluding Remarks

The main objective of this chapter was to present three different approaches that can be used for designing unknown input observers for non-linear discrete-time systems. In particular, Section 4.1 introduced the concept of an extended unknown input observer. It was shown, with the help of the Lyapunov method, that such a linerarisation-based technique is convergent under certain conditions. However, it was pointed out that these conditions were obtained under a very restrictive assumption. To tackle this task, a novel structure and design procedure of the EUIO were proposed in Section 4.2. As a result, convergence conditions were obtained which do not require restrictive assumptions, as was the case in Section 4.1. It was also shown that the achieved convergence condition is less restrictive than the one resulting from the approach described in [19]. The common advantage of the approaches presented in Sections 4.1 and 4.2 is related to their implementation simplicity. Indeed, the design procedures are almost as straightforward as their counterparts for linear systems. Minor modifications that significantly improve the convergence and the convergence rate are also very easy to implement.

The third observer structure presented in this chapter (see Section 4.3) was dedicated to the discrete-time Lipschitz system. In particular, the main objective was to develop efficient approaches to designing observers for discrete-time Lipschitz non-linear systems. In particular, with the use of the Lyapunov method, three different convergence criteria were developed. The difference between these criteria lies in the way the Lyapunov function is calculated. All these techniques introduce a level of conservatism related to the relative inaccuracy of a given

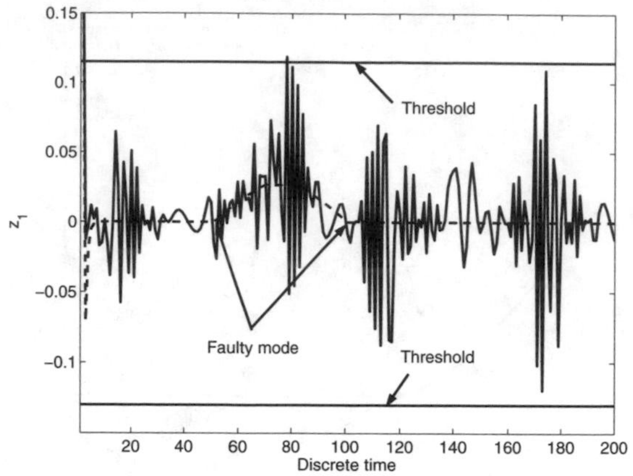


**Fig. 4.8.** Residual $z_{1,k} = y_{1,k} - \hat{y}_{1,k}$ obtained with the UIO (dashed line) and the conventional observer

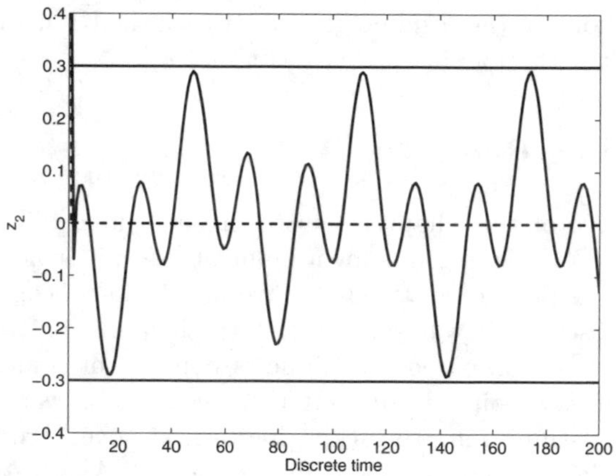


**Fig. 4.9.** Residual $z_{2,k} = y_{2,k} - \hat{y}_{2,k}$ obtained with the UIO (dashed line) and the conventional observer

technique. Based on the achieved results, three different design procedures were proposed. These procedures were developed in such a way that the design problem boils down to solving a set of linear matrix inequalities or solving the generalised eigenvalue minimisation problem under LMI constraints. Experimental results confirm the effectiveness of the proposed design procedures. In particular, it was shown that the proposed approach can be effectively applied to design an observer for a flexible link robot, which is a frequently used benchmark for observers for Lipschitz systems. Moreover, the convergence rate provided by the proposed observer is significantly higher than the one obtained with the techniques present in the literature.

Another objective was to show how to apply the proposed techniques to systems with unknown inputs. This was realised with the use of a suitable system transformation that converts system description with an unknown input into a system description without it. Experimental results confirm the effectiveness of the proposed design procedures and show the potential profits that can be achieved while applying the proposed approach to the FDI scheme.

# 5. Parameter Estimation-Based FDI

The approaches presented in the preceding part of this book do not employ an additional design freedom that can be achieved by suitably scheduling input signals. Thus, from this point of view the presented approaches can be perceived as Passive Fault Diagnosis (PFD) tools. There is no doubt that it is profitable to design the input signal in such a way as to improve FDI in some sense and to achieve the control goal at the same time. The design of such an input signal is usually based on the minimisation of an additional criterion (e.g., the minimisation of fault detection delay). An important question is how to apply this input signal to the observed system. Some extra inputs which are not used for control can be utilised to excite the system [196]. This is not a common case, and an input signal is more often applied through the standard inputs in the so-called test period [24, 40]. The test period must be short and wisely chosen because no control is generated. This problem can be solved using an input signal which fulfils two opposed aims: excitation and control.

The area of Active Fault Diagnosis (AFD) has received considerable attention during the recent years. AFD has been considered in a number of papers (see, e.g., [24, 124] and the reference therein). AFD will, in general, result in a faster fault detection than the one obtained with PFD.

Unfortunately, the main drawback of the above-mentioned approaches is the fact that they are designed for linear systems and it is extremely difficult to use them for non-linear ones. Thus, the only feasible way is to use an off-line input design procedure that can be used for non-linear systems [7, 50, 167, 170]. The above mentioned procedure is primarily employed for parameter estimation but its incontestable appeal is that it reduces the resulting model uncertainty. This feature will be exploited in the subsequent part of this chapter. Apart from designing an optimal input sequence, the above-described experimental design strategy [7, 50, 167, 170] can also be used for determining the optimal values of all variables having some influence on the parameter estimation process (this feature will be clearly illustrated in Section 5.2).

M. Witczak: Model. and Estim. Strat. for Fault Diagn. of Non-Linear Syst. LNCIS 354, pp. 85–101, 2007.
springerlink.com 

At this point, it should by pointed out that the approaches presented in this chapter are restricted to static non-linear systems. Experimental design for dynamic non-linear systems is extremely complicated and there is no general solution that can be applied to a wide class of non-linear systems. For a comprehensive treatment regarding this subject the reader is referred to [167].

The chapter is organised as follows: Section 5.1 present an elementary background regarding experimental design for non-linear systems, while Section 5.2 presents a comprehensive case study regarding impedance measurement and fault detection with experimental design strategies. It should be pointed out that some of the results presented in Section 5.2 were originally presented in [180].

## 5.1 Experimental Design

It is well known that data collection requires some effort that can be expressed in the form of time, financial means, or other material resources. On the other hand, a proper design makes it possible to exploit the available resources in the most efficient way. The purpose of experimental design is to determine experimental conditions that make it possible to determine the model parameters as accurately as possible, i.e., with a possibly small model uncertainty. As was mentioned in Section 2.3, the measurements used for parameter estimation are corrupted by noise and disturbances, which contribute directly to parameter uncertainty. As a result, there is no unique $\hat{\boldsymbol{p}}$ that is consistent with a given set of measurements, but there is a parameter set $\mathbb{P}$ that satisfies this requirements. Thus, the main objective is to design an experiment in such a way as to minimise the confidence region $\mathbb{P}$. It is an obvious fact that the smaller the confidence region, the more sensitive the fault diagnosis scheme. This means that experimental design is of paramount importance for parameter estimation-based FDI.

To explain this in a more formal way [170], let us assume that the $i$th output observation is $y(\boldsymbol{\xi}^i)$, where $\boldsymbol{\xi}^i$ is a vector describing the experimental conditions (e.g., measurement time, the shape and value of inputs, etc.) under which the $i$th measurement is to be collected. When $n_t$ observations are collected, then the concatenation of $\boldsymbol{\xi}^i$, $i = 1, \ldots, n_t$ gives $\boldsymbol{\xi} = [(\boldsymbol{\xi}^1)^T, \ldots, (\boldsymbol{\xi}^{n_t})^T]^T$, which characterises all experimental conditions to be optimised. Let $\mathbb{X}$ denote the set of all feasible values of $\boldsymbol{\xi}$. Then the optimal experimental design can be formulated as a constrained optimisation problem of the form

$$\boldsymbol{\xi}^* = \arg\min_{\boldsymbol{\xi} \in \mathbb{X}} \Phi(\boldsymbol{\xi}), \tag{5.1}$$

where $\Phi(\cdot)$ stands for the scalar cost function, which can be defined in many different ways [7, 50, 167, 170]. However, the most common approach is to use $\Phi(\cdot)$, which is related to the so-called Fisher Information Matrix (FIM) [7, 50, 167, 170]. A valuable property of the FIM is that its inverse constitutes an approximation of the covariance matrix for the estimates of the system parameters $\boldsymbol{p}$ [66].

To express this in a more formal way, let us start with a simple linear system described by

$$y_k = \boldsymbol{r}_k^T \boldsymbol{p} + v_k, \tag{5.2}$$

where $\boldsymbol{r}_k$ depends on the $k$th experimental condition $\boldsymbol{\xi}^k$, and $v$ stands for the zero-mean, uncorrelated, Gaussian noise sequence. Thus, for a set of $n_t$ observations the FIM can be expressed by

$$\boldsymbol{P}^{-1} = \sum_{k=1}^{n_t} \boldsymbol{r}_k \boldsymbol{r}_k^T. \tag{5.3}$$

Since the FIM is given, then it is possible to describe the most frequently used cost functions $\Phi(\cdot)$ that can be used for solving (5.1). These cost functions are usually related to the following criteria [7, 50, 167, 170]:

- D-optimality criterion:

$$\Phi(\xi) = \det \boldsymbol{P}; \tag{5.4}$$

- E-optimality criterion ($\lambda_{\max}(\cdot)$ stands for the maximum eigenvalue of its argument):

$$\Phi(\xi) = \lambda_{\max}(\boldsymbol{P}); \tag{5.5}$$

- A-optimality criterion:

$$\Phi(\xi) = \operatorname{trace} \boldsymbol{P}. \tag{5.6}$$

Thus, a D-optimum design minimises the volume of the confidence ellipsoid describing the feasible parameter set $\mathbb{P}$ (see, e.g., [7, Sec. 6.2] for further explanations). An E-optimum design minimises the length of the largest axis of the same ellipsoid. An A-optimum design suppresses the average variance of parameter estimates.

It is clear that a direct optimisation of (5.1) is possible in relatively simple cases only. Such a design study is to be considered in Section 5.2. In other situations, specialised algorithms such as the Wynn-Fedorov algorithm [7, 50, 167, 170] have to be applied. Such an algorithm is to be employed in Section 7.1.

Another problem is related to non-linear systems which can be described by

$$y_k = g(\boldsymbol{\xi}^k, \boldsymbol{p}) + v_k. \tag{5.7}$$

In this case, the FIM is given by (5.3) with

$$\boldsymbol{r}_k = \frac{\partial g(\boldsymbol{\xi}^k, \boldsymbol{p})}{\partial \boldsymbol{p}}. \tag{5.8}$$

It is important to note that (5.8) depends on the values of the unknown parameters $\boldsymbol{p}$. Thus, in most practical situations some estimates of $\boldsymbol{p}$ are used for calculating the FIM, i.e., it is given by (5.3) with

$$\boldsymbol{r}_k = \left.\frac{\partial g(\boldsymbol{\xi}^k, \boldsymbol{p})}{\partial \boldsymbol{p}}\right|_{p=\hat{\boldsymbol{p}}}. \tag{5.9}$$

Thus, the more accurate estimates, the better the obtained experimental design.

Since all elementary background on experimental design is given, then it is possible to show all the advantages that can be gained while using it for developing FDI schemes. To tackle this problem, an example regarding impedance measurement and diagnosis is to be employed and detailed in Section 5.2.

## 5.2 Impedance Measurement and Diagnosis

Various physical quantities like, e.g., force, pressure, flow and displacement can be transformed into an impedance through different transducers. The measurements collected in such a way constitute the basic source of knowledge regarding any industrial system being controlled or diagnosed [96, 179]. Many different techniques and equipment have been developed over the past years so as to meet demands on different operational ranges, accuracies, measurement rates, specific application targets, and costs. So far, there has been no doubt that high-accuracy measurements can be achieved by using AC bridges that are balanced either manually or iteratively. Unfortunately, such a balancing technique excludes fast measurement rates that are required by modern control and fault diagnosis systems [96, 179]. In order to settle such a challenging problem, many different computer-based AC bridges have been developed over the last fifteen years. Dutta *et al.* [44] introduced the idea of a "virtual AC bridge" with a virtual arm implemented with the help of a microcomputer. Awad *et al.* [8] made an important contribution to this approach with respect to its convergence. Recently, Dutta *et al.* [45] proposed another modification of this impedance measurement technique. In all of the above papers, the authors formulated the impedance measurement problem as the non-linear-in-parameter estimation [170] one. To solve such a problem, they employed the gradient descent algorithm [170], which made the required measurement rate difficult to attain. To settle this problem, Kaczmarek *et al.* [87] employed the so-called bounded-error parameter estimation technique [170]. In spite of the high convergence rate, the technique does not make it possible to attain accuracy comparable with [8, 45]. A slightly different approach was proposed by Angrisani *et al.* [5]. This technique is based on a bridge-balance loop comprising two internal loops. The first one is used for tuning the capacitive (inductive) part of the impedance, while the second one for arranging the resistive part. The main drawback of this approach is that fast convergence seems very hard to attain. A fast balancing bridge was proposed by Zhang *et al.* [195], and it seems suitable for a wide range of industrial applications.

Unfortunately, the researchers have not considered the relation between experimental conditions [167, 170] and the accuracy of the measurements. In this work, experimental conditions are related to optimum selection of the reference resistance of the bridge and the sampling time. Angrisani *et al.* [5] and Zhang *et al.* [195] discovered, by numerous simulations, that the closer the selected reference resistance is to the measured impedance, the more precise measurement results can be obtained. However, they did not provide any explicit formulae determining the reference resistance.

In this section, the answer to the challenging problem of optimum experimental conditions is provided, i.e., explicit formulae for the reference resistance and the sampling time are developed. The subsequent part of this section is organised as follows: Section 5.2.1 shows the formulation of the problem and the proposed measurement technique. In Section 5.2.2, the results regarding optimum experimental conditions are provided. Section 5.2.3 shows a way of extending the proposed approach to inductive impedances. Section 5.2.4 is devoted to robustness problems, which are very important in practical applications. Finally, section 5.2.5 presents experimental results that confirm the effectiveness of the proposed approach.

### 5.2.1 Problem Formulation

The objective of this section is to propose a new impedance measurement scheme. This work is motivated by the approach presented in [87]. The virtual bridge is composed of two arms, namely, a real (hardware) arm, as shown in Fig. 5.1, and a virtual arm, implemented with the help of a computer. As has already been mentioned, there is no literature providing any analytical rules that can be used for obtaining the reference resistance $R_r$ and the sampling time. Another question that arises while analysing [8, 44] is as follows: Is it really necessary to use non-linear parameter estimation techniques for estimating $R$ and $C$? First, let us observe that for the scheme presented in Fig. 5.1 the following current equality can be established:

$$C\frac{dy_M(t)}{dt} + \frac{y_M(t)}{R} = \frac{u(t) - y_M(t)}{R_r}. \tag{5.10}$$

Assuming that $u(t) = U\sqrt{2}\sin(\omega t)$, the steady-state solution of (5.10) can be written as

$$y_M(t) = \rho U\sqrt{2}R((R + Rr)\sin(\omega t) - R_r RC\omega\cos(\omega t)), \tag{5.11}$$

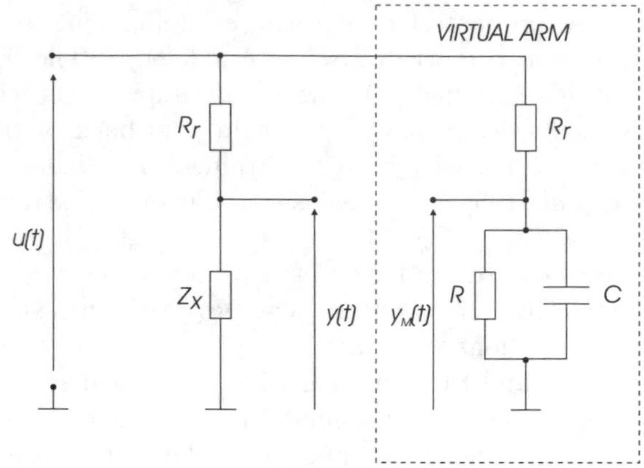


**Fig. 5.1.** Impedance measurement scheme

where $\rho = \left(R^2 + 2R_rR + R_r^2(1+\omega^2R^2C^2)\right)^{-1}$. The equation (5.11) can be transformed into a discrete-time form and written as follows:

$$y_{M,k} = p_1u_{1,k} + p_2u_{2,k}, \tag{5.12}$$

with

$$p_1 = \rho R(R + R_r), \quad p_2 = \rho R_r C\omega R^2, \tag{5.13}$$

and $u_{1,k} = U\sqrt{2}\sin(\omega k\tau)$, $u_{2,k} = U\sqrt{2}\cos(\omega k\tau)$, where $\tau$ stands for the sampling time. In this work, it is assumed that $u_{1,k}$ and $u_{2,k}$ are available. This is a mild assumption since it is not difficult to design hardware providing such signals. Another important fact that can be observed while analysing (5.12) is that it can be perceived as a linear-in-parameter model with respect to $p_1$ and $p_2$. Contrary to [8, 44], where non-linear parameter estimation techniques were employed for obtaining $R$ and $C$, it is proposed to use the classic Recursive Least-Square (RLS) algorithm [170] for theestimation of $p_1$ and $p_2$. Such an algorithm can be written as follows:

$$\hat{\boldsymbol{p}}_k = \hat{\boldsymbol{p}}_{k-1} + \boldsymbol{k}_k\varepsilon_k, \tag{5.14}$$

$$\boldsymbol{k}_k = \boldsymbol{P}_{k-1}\boldsymbol{r}_k\left(1 + \boldsymbol{r}_k^T\boldsymbol{P}_{k-1}\boldsymbol{r}_k\right)^{-1}, \tag{5.15}$$

$$\varepsilon_k = y_k - g(\hat{\boldsymbol{p}}_{k-1}, \boldsymbol{u}_k), \tag{5.16}$$

$$\boldsymbol{P}_k = \left[\boldsymbol{I}_{n_p} - \boldsymbol{k}_k\boldsymbol{r}_k^T\right]\boldsymbol{P}_{k-1}, \tag{5.17}$$

$g(\hat{\boldsymbol{p}}_k, \boldsymbol{u}_{k+1}) = \hat{p}_{1,k}u_{1,k+1} + \hat{p}_{2,k}u_{2,k+1}$, $\boldsymbol{r}_k = [u_{1,k}, u_{2,k}]^T$, $\hat{\boldsymbol{p}}_k = [\hat{p}_{1,k}, \hat{p}_{2,k}]^T \in \mathbb{R}^{n_p}$ denotes the $k$th estimate of $\boldsymbol{p}$, and $k = 1, \ldots, n_t$. Thus, knowing $\hat{\boldsymbol{p}}$ it is possible to obtain the estimates of $R$ and $C$ according to the following equations:

$$\hat{R} = -\frac{R_r(\hat{p}_1^2 + \hat{p}_2^2)}{\hat{p}_1^2 + \hat{p}_2^2 - \hat{p}_1}, \tag{5.18}$$

$$\hat{C} = -\frac{\hat{p}_2}{R_r\omega(\hat{p}_1^2 + \hat{p}_2^2)}, \tag{5.19}$$

obtained by solving (5.13) with respect to $R$ and $C$. It should also be pointed out that when there is no need for on-line estimation of the impedance, then the classic, non-recursive least-square algorithm can be employed. The well-known advantage of this algorithm, comparing with its recursive counterpart, is that the highest estimation accuracy can be attained with smaller $n_t$. In this case, the estimates of $p_1$ and $p_2$ can be computed as follows:

$$\hat{p}_1 = \frac{\gamma_2\eta - \beta_2\gamma_1}{\eta^2 - \beta_1\beta_2}, \quad \hat{p}_2 = \frac{\gamma_1\eta - \beta_1\gamma_2}{\eta^2 - \beta_1\beta_2}, \tag{5.20}$$

where

$$\gamma_i = \sum_{k=1}^{n_t} u_{i,k}y_k, \quad \eta = \sum_{k=1}^{n_t} u_{1,k}u_{2,k}, \quad \beta_i = \sum_{k=1}^{n_t} u_{i,k}^2. \tag{5.21}$$

As can be found in the literature [170] regarding the RLS algorithm, the initial matrix $\boldsymbol{P}_k$, i.e., $\boldsymbol{P}_0$ should be set as $\boldsymbol{P}_0 = \gamma\boldsymbol{I}$, where $\gamma$ stands for a sufficiently large positive constant (usually $10^3$–$10^{20}$). When some rough values of $R$ and $C$ are known then $\hat{\boldsymbol{p}}_0$ should be initialised according to (5.13). Otherwise, it can be observed from (5.18) that $\hat{p}_1^2 + \hat{p}_2^2 - \hat{p}_1 < 0$ and hence

$$\frac{1}{2} - \frac{1}{2}\sqrt{1 - 4\hat{p}_2^2} < \hat{p}_1 < \frac{1}{2} + \frac{1}{2}\sqrt{1 - 4\hat{p}_2^2}. \tag{5.22}$$

Since $\hat{p}_2$ should satisfy $1 - 4\hat{p}_2^2 > 0$ and (5.19) indicates that $\hat{p}_2 < 0$, then it is clear that

$$-\frac{1}{2} < \hat{p}_2 < 0. \tag{5.23}$$

Thus, when no knowledge is available about $R$ and $C$, then $\hat{\boldsymbol{p}}_0$ should be set so as to satisfy (5.22)–(5.23).

### 5.2.2 Experimental Design

One objective of this section is to provide rules for computing the accuracy of the measured impedance, i.e., a set of all $R$ and $C$ that are consistent with the measurements of $\boldsymbol{u}$ and $y$. The main objective is to provide optimum experimental conditions, i.e., explicit formulae for the reference resistance and the sampling time are developed that make it possible to increase measurement accuracy. The proposed solution is based on the following assumption:

$$y_k = y_{M,k} + v_k, \tag{5.24}$$

while $v$ stands for the zero-mean, uncorrelated, Gaussian noise sequence. In other words, $v_k$ represents the difference between the output of the model (5.12) and $y_k$ (cf. Fig. 5.1).

### Confidence region and fault detection

Since estimates of $R$ and $C$ can be obtained according to (5.18)–(5.19), the next problem being considered is to obtain a set of all possible $R$ and $C$ that are consistent with the measurements of $\boldsymbol{u}$ and $y$. Such a set can be obtained with the use of the $(1-\alpha)100\%$ confidence region [170] for $\boldsymbol{p}$ and the equations (5.13), where $\alpha$ stands for the significance level. As a result, the following inequality is given:

$$\boldsymbol{s}_k^T \boldsymbol{P}_k^{-1} \boldsymbol{s}_k \leq 2\hat{\sigma}_k^2 F_{\alpha,2,k-2} \tag{5.25}$$

where

$$\boldsymbol{s}_k = \hat{\boldsymbol{p}}_k - \rho \left[R(R+R_r), R_r C \omega R^2\right]^T, \tag{5.26}$$

and $F_{\alpha,2,k-2}$ stands for the F-Snedecor distribution quantile with 2 and $k-2$ degrees of freedom, and $\hat{\sigma}$ is the estimate of the standard deviation of $v$. The inequality (5.25) is very important from the point of view of fault detection and control of industrial systems [96, 179]. Indeed, it can be used for checking that the measured impedance satisfies the predefined bounds. On the other hand, the problem of fault detection can be transformed into the task of hypotheses testing. This means that, at the $\alpha$-level, the hypothesis

$$\begin{aligned} &\mathcal{H}_0 : \ (R, C) = (R_0, C_0) \\ &\text{vs.} \\ &\mathcal{H}_1 : \ (R, C) \neq (R_0, C_0), \end{aligned} \tag{5.27}$$

whereas $R_0$, $C_0$ are the required (nominal) values of $R$ and $C$, is rejected when the inequality (5.25) is violated. The acceptance of the hypothesis $\mathcal{H}_1$ denotes faulty behavior of the impedance.

### Optimum experimental conditions

As can be seen from (5.25), the size of the confidence region depends on the FIM, $\boldsymbol{P}^{-1}$. Thus, optimal experimental conditions can be obtained by optimising some scalar function $\Phi(\boldsymbol{P}^{-1})$. As has already been mentioned, such a function can be defined in several different ways [167, 170]. In this section, the so-called *D-optimality criterion* is used. This means that an appropriate selection of experimental conditions will make it possible to obtain a more reliable fault diagnosis system (through more accurate measurements of the impedance) than those designed without it [40]. It should also be pointed out that experimental conditions are developed for $R$ and $C$ but not for $p_1$ and $p_2$. This means that all dependencies among $R_r$, $\omega$, $\tau$, $R$, and $C$ that provide an additional source of knowledge are exploited. First, let us define the FIM:

$$\boldsymbol{P}^{-1} = \sum_{k=1}^{n_t} \boldsymbol{r}_k \boldsymbol{r}_k^T, \quad \boldsymbol{r}_k = \left[\frac{\partial y_{M,k}}{\partial R}, \frac{\partial y_{M,k}}{\partial C}\right]^T. \tag{5.28}$$

The purpose of further deliberations is to obtain D-optimum values of $R_r$ and $\tau$, i.e., $R_r$ and $\tau$ that maximise $\det\left(\boldsymbol{P}_k^{-1}\right)$.

It can be observed that

$$\begin{aligned}
\boldsymbol{r}_k &= \boldsymbol{P}_1 \boldsymbol{r}_{1,k}, \quad \boldsymbol{P}_1 = \sqrt{2} U R_r \rho^2 \mathrm{diag}(1, \omega R^2), \\
\boldsymbol{r}_{1,k} &= [a \sin(\omega k \tau) + b \cos(\omega k \tau), b \sin(\omega k \tau) - a \ \cos(\omega k \tau)], \\
a &= R^2 + 2R_r R + R_r^2(1 - \omega^2 R^2 C^2), \quad b = -2C\omega(R_r R^2 + R R_r^2).
\end{aligned}$$

Bearing in mind that the fact that

$$\begin{aligned}
&\sqrt{a^2+b^2}\sin(\omega k \tau + \arctan(a/b)) = a \sin(\omega k \tau) + b \cos(\omega k \tau), \\
&\sqrt{a^2+b^2} = \rho^{-1}, \qquad (5.29)
\end{aligned}$$

it is possible to write:

$$\begin{aligned}
\boldsymbol{r}_k &= \boldsymbol{P}_2 \boldsymbol{r}_{2,k}, \quad \boldsymbol{P}_2 = \sqrt{2} U R_r \rho \mathrm{diag}(1, \omega R^2), \\
\boldsymbol{r}_{2,k} &= [\sin(\omega k \tau + \arctan(a/b)), \sin(\omega k \tau + \arctan(-b/a))]^T . \qquad (5.30)
\end{aligned}$$

Using the equations (5.30), the FIM can be given as follows:

$$\boldsymbol{P}^{-1} = \boldsymbol{P}_2 \sum_{k=1}^{n_t} \boldsymbol{r}_{2,k} \boldsymbol{r}_{2,k}^T \boldsymbol{P}_2. \qquad (5.31)$$

The main difficulty associated with further deliberations is concerned with the selection of the number of measurements $n_t$. Indeed, it is very difficult to give $n_t$ a priori. In order to perform further derivations, two relatively non-restrictive assumptions are formulated:

**Assumption 5.1.** Sampling starts exactly at the beginning of the period of $u(t)$.

**Assumption 5.2.** The ratio between the period of $u(t)$ and the sampling interval is a rational number.

Under the above assumptions and due to the nature of $\sin(\omega k \tau)$, it is easy to see that experimental conditions are cyclically repeated. When some experiments are repeated, then the number $n_e$ of distinct experimental conditions is smaller than the total number of observations $n_t$. The design resulting from this approach is called the continuous experimental design [167, 170]. The FIM can then be defined as

$$\boldsymbol{P}^{-1} = \boldsymbol{P}_2 \sum_{k=1}^{n_e} \mu_k \boldsymbol{r}_{2,k} \boldsymbol{r}_{2,k}^T \boldsymbol{P}_2. \qquad (5.32)$$

$\mu_1, \ldots, \mu_{n_e}, \mu_k \in [0,1]$ are perceived as weights associated with the $k = 1, \ldots, n_e$ experimental conditions, which satisfy $\sum_{k=1}^{n_e} \mu_k = 1$. Caratheodory's theorem then indicates that (5.32) can be written with a linear combination of at most $n_e = n_p(n_p+1)/2+1$ ($n_e = 4$ since we have two parameters $R$ and $C$) matrices $\boldsymbol{r}_{2,k}\boldsymbol{r}_{2,k}^T$. In the sequel, the setting $n_e = 4$ is employed.

Using (5.32), it can be shown that

$$\det\left(\boldsymbol{P}^{-1}\right) = \det\left(\boldsymbol{P}_2\right)^2 \det\left(\sum_{k=1}^{n_e} \mu_k \boldsymbol{r}_{2,k} \boldsymbol{r}_{2,k}^T\right).$$

After some relatively easy but lengthy calculations, it can be shown that

$$\det\left(\sum_{k=1}^{n_e} \mu_k \boldsymbol{r}_{2,k} \boldsymbol{r}_{2,k}^T\right) = \sin(\omega\tau)^2 \left(16\mu_1\mu_4 \cos(\omega\tau)^4 \right.$$
$$\left. +4(\mu_1\mu_3 + \mu_2\mu_4 - 2\mu_1\mu_4)\cos(\omega\tau)^2 + \mu_1\mu_4 + \mu_1\mu_2 + \mu_2\mu_3 + \mu_3\mu_4\right) \quad (5.33)$$

It can be observed that (5.33) is independent of $R$, $C$ and $R_r$. On the other hand, $\boldsymbol{P}_2$ does not depend on $\tau$. This means that the maximisation of the determinant of the FIM (or the minimization of its inverse) with respect to $\tau$ is equivalent to

$$\tau^* = \arg \max_{\tau>0,\mu_i,i=1,\dots,n_e} \det\left(\sum_{k=1}^{n_e} \mu_k \boldsymbol{r}_{2,k} \boldsymbol{r}_{2,k}^T\right), \quad (5.34)$$

while the maximisation of the determinant of the FIM with respect to $R_r$ is equivalent to

$$R_r^* = \arg \max_{R_r>0} \det(\boldsymbol{P}_2) = \arg \max_{R_r>0} 2\rho^2\omega U^2 R^2 R_r^2. \quad (5.35)$$

The solution of (5.34) is given as follows:

$$\tau^* = \frac{\pi(1+i)}{2\omega}, \quad i = 0, \quad (5.36)$$

with $\mu_k = 1/4$, $k = 1, \dots, n_e = 4$. Note that $i$ is not greater than zero, which corresponds to the sampling frequency more than two times larger than the frequency of the input signal. Finally, the D-optimum value of the reference resistance $R_r^*$ (being the solution of (5.35)) can be written according to

$$R_r^* = \frac{R}{\sqrt{1+\omega^2 R^2 C^2}}. \quad (5.37)$$

## Other properties

The objective of this section is to investigate the influence of the experimental conditions (5.36) and (5.37) on the estimation accuracy of $\boldsymbol{p}$. First, let us define the FIM for $\boldsymbol{p}$:

$$\boldsymbol{P}^{-1} = \sum_{k=1}^{n_t} \boldsymbol{r}_k \boldsymbol{r}_k^T,$$
$$\boldsymbol{r}_k = \left[\frac{\partial y_{M,k}}{\partial p_1}, \frac{\partial y_{M,k}}{\partial p_2}\right]^T = U\sqrt{2}\left[\sin(\omega\tau k), \cos(\omega\tau k)\right]^T. \quad (5.38)$$

Similarly as in Section 5.2.2, the FIM for continuous design can be defined as

$$\boldsymbol{P}^{-1} = \sum_{k=1}^{n_e=4} \mu_k \boldsymbol{r}_k \boldsymbol{r}_k^T. \tag{5.39}$$

Substituting (5.36) into (5.39), it can be shown that

$$\boldsymbol{P}^{-1} = 2U^2 \mathrm{diag}(\mu_1 + \mu_3, \mu_2 + \mu_4). \tag{5.40}$$

From (5.40), it can be observed that the FIM is diagonal. A design satisfying this property is called the *orthogonal design.* Its appealing property is that the covariance between the parameters $p_1$ and $p_2$ equals zero, which means that they are estimated independently. The remaining task is to check if the experimental conditions (5.36)–(5.37) are D-optimum for $\boldsymbol{p}$. In order to do that, the following useful criterion can be used [167, 170]:

$$\boldsymbol{r}_k^T \boldsymbol{P} \boldsymbol{r}_k \leq n_p \tag{5.41}$$

when the equality holds for $\boldsymbol{r}_k$ satisfying the experimental conditions (5.36) and (5.37). Substituting $n_p = 2$ and then (5.40) into (5.41), it can be shown that

$$\frac{\sin(\frac{1}{2}\pi k)^2}{\mu_1 + \mu_3} + \frac{\cos(\frac{1}{2}\pi k)^2}{\mu_2 + \mu_4} \leq 2. \tag{5.42}$$

Setting $\mu_k = 1/4,\ k = 1, \ldots, n_e = 4$ in (5.42) implies that the experimental design (5.36)–(5.37) is D-optimum and orthogonal for $\boldsymbol{p}$, i.e., the FIM is a diagonal matrix.

### 5.2.3 Inductive Impedance

The main objective of this section is to derive D-optimum experimental conditions for an inductive impedance. This sections presents the main results only that are obtained according to the derivation presented in the preceding sections. First, let us observe that for an inductive impedance the following current equality can be established:

$$\frac{y_M(t)}{R} + \frac{1}{L}\int_0^t y_M(t)\mathrm{d}t = \frac{u(t) - y_M(t)}{R_r}. \tag{5.43}$$

The discrete-time steady-state solution of (5.43) can be written as

$$y_{M,k} = p_1 u_{1,k} + p_2 u_{2,k}, \tag{5.44}$$

where

$$p_1 = \rho L^2 \omega^2 (R_r + R) R, \quad p_2 = \rho L \omega R_r R^2, \tag{5.45}$$

where $\rho = \left((R+R_r)^2\omega^2L^2 + R^2R_r^2\right)^{-1}$. Knowing $\hat{\boldsymbol{p}}$, it is possible to obtain the estimates of $R$ and $L$ according to the following equations:

$$\hat{R} = -\frac{R_r(\hat{p}_1^2 + \hat{p}_2^2)}{\hat{p}_1^2 + \hat{p}_2^2 - \hat{p}_1}, \tag{5.46}$$

$$\hat{L} = \frac{R_r(\hat{p}_2^2 + \hat{p}_1^2)}{\hat{p}_2\omega}, \tag{5.47}$$

obtained by solving (5.45) with respect to $R$ and $L$. Finally, the D-optimum sampling time is given by (5.36) while the D-optimum reference resistance is

$$R_r^* = \frac{RL\omega}{\sqrt{R^2 + \omega^2L^2}}. \tag{5.48}$$

Moreover, the confidence region (5.25) can be computed by using

$$\boldsymbol{s}_k = \hat{\boldsymbol{p}}_k - \rho\left[L^2\omega^2R(R+R_r), L\omega R_r\omega R^2\right]^T. \tag{5.49}$$

Similarly as in Section 5.2.1, the initial values of the parameters $p_1$ and $p_2$ should satisfy

$$\frac{1}{2} - \frac{1}{2}\sqrt{1-4\hat{p}_2^2} < \hat{p}_1 < \frac{1}{2} + \frac{1}{2}\sqrt{1-4\hat{p}_2^2}, \tag{5.50}$$

and

$$0 < \hat{p}_2 < \frac{1}{2}. \tag{5.51}$$

Another important property is that the optimality conditions for $p_1$ and $p_2$ are valid for inductive impedances as well.

### 5.2.4 Towards Robustness

It is clear from (5.37) (and (5.48)) that the D-optimum value of the reference resistance $R_r$ depends on the values of the unknown parameters $R$ and $C(L)$. Indeed, it is well known from the literature [167, 170] that the dependence on the parameters that enter non-linearly into the model is an unappealing property of the non-linear optimum experimental design. One way out of this problem is to use the so-called sequential design [167, 170]. When some rough estimates of $R$ and $C(L)$, then $R_r$ can be calculated according to (5.37) (or (5.48)) and the impedance measurement procedure can be started. As a result, a more accurate impedance estimate can be obtained which can be employed to find new $R_r$. This two-step sequential procedure can be repeated several times until satisfactory results are obtained, i.e., a suitable measurement accuracy is accomplished. On the other hand, when some prior bounds for $R$ and $C(L)$ are given, i.e., $R \in [R_{\min}, R_{\max}]$ and $C(L) \in [C(L)_{\min}, C(L)_{\max}]$, then it is possible to use (5.37) (or (5.48)) to compute $R_r$ for the average values of $R$ and $C(L)$ defined as $\bar{C}(\bar{L}) = 0.5(C(L)_{\min} + C(L)_{\max})$, $\bar{R} = 0.5(R_{\min} + R_{\max})$ (see [167, 170] for further comments about average D-optimality).

### 5.2.5 Simulation Results

The main objective of this section is to perform a computer simulation-based analysis of the impedance measurement technique presented in the preceding sections. Only the results for capacitive impedances are presented (many computer simulations show that similar results can be achieved for inductive impedances measured according to the approach presented in Section 5.2.3).

In all numerical experiments, measured $y_k$ was generated according to

$$y_k = y_{M,k} + v_k, \tag{5.52}$$

and $v_k \sim \mathcal{U}(-2\delta_q, 2\delta_q)$, where $\mathcal{U}$ stands for the uniform distribution, and $\delta_q$ is the quantisation error of the 12-bit ADC defined as

$$\delta_q = \frac{1}{2}\frac{U}{2^{12}-1}, \tag{5.53}$$

where $U = 2.5[\mathrm{V}]$. In all experiments, the input signal frequency was $1[\mathrm{kHz}]$, the sampling time $\tau$ was selected according to (5.36), $\boldsymbol{p}_0 = [-0.1, -0.1]^T$, and the number of measurements of $y_k$ was $n_t = 4000$.

#### Reference resistance, confidence region and fault detection

Let us consider a numerical simulation example for the following parameters: $R = 10[\mathrm{k}\Omega]$, $C = 540[\mathrm{pF}]$. Two different experiments were performed for two different values of $R_r$, i.e., $R_r = R_r^*$ (cf. (5.37)) and $R_r = 1[\mathrm{k}\Omega]$. Each of the above experiments was repeated 500 times. Figures 5.2 and 5.3 show the histograms of the relative measurements errors:

$$\delta_R = \frac{R-\hat{R}}{R}100[\%], \quad \delta_C = \frac{C-\hat{C}}{C}100[\%], \tag{5.54}$$

for $R_r = R_r^*$ and $R_r = 1[\mathrm{k}\Omega]$, respectively. From these results, it can be seen that a considerable increase in measurement accuracy can be achieved with D-optimum experimental conditions.

The purpose of the subsequent example is to use the approach developed in Section 5.2.2 for fault detection of an impedance. Let us consider the following parameter set: $R = 500[\Omega]$, $C = 0.75[\mathrm{nF}]$. Two different experiments were performed for two different values of $R_r$, i.e., $R_r = R_r^*$ (cf. (5.37)) and $R_r = 5[\mathrm{k}\Omega]$. Let us assume that non-faulty $R$ and $C$ are $R = 500.03[\Omega]$ and $C = 0.76[\mathrm{nF}]$. Thus, the problem of fault detection boils down to the task of testing

$$\begin{aligned} &\mathcal{H}_0: \ (R,C) = (500.03[\Omega], 0.76[\mathrm{nF}]),\\ &\text{vs.}\\ &\mathcal{H}_1: \ (R,C) \neq (500.03[\Omega], 0.76[\mathrm{nF}]). \end{aligned} \tag{5.55}$$

Figure 5.4 shows the confidence region (5.25) of $R$ and $C$ for $R_r = 5[\mathrm{k}\Omega]$ (larger) and $R_r = R_r^*$, respectively (assuming $\alpha = 0.01$, i.e., the 99% confidence region).

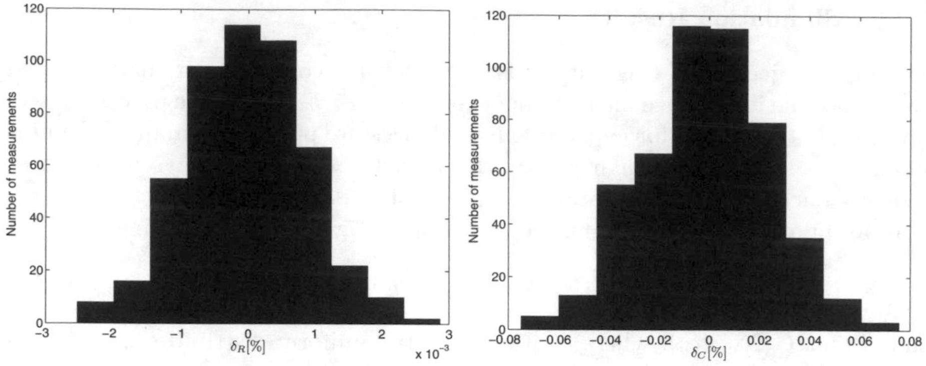


**Fig. 5.2.** Relative errors for $R_r = R_r^*$

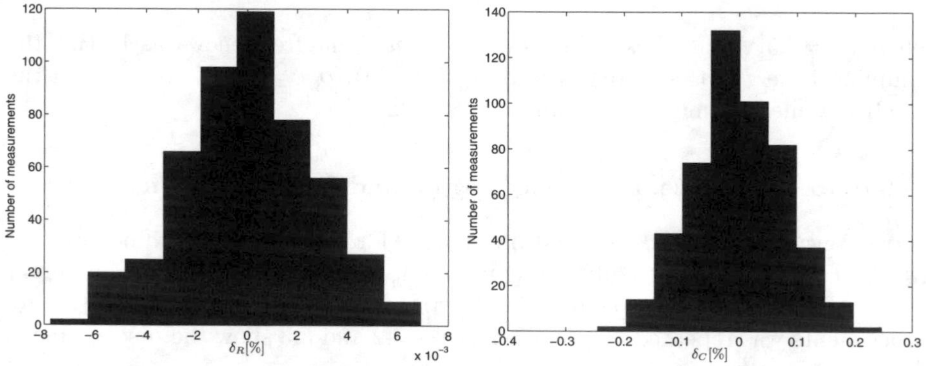


**Fig. 5.3.** Relative errors for $R_r = 1[\mathrm{k}\Omega]$

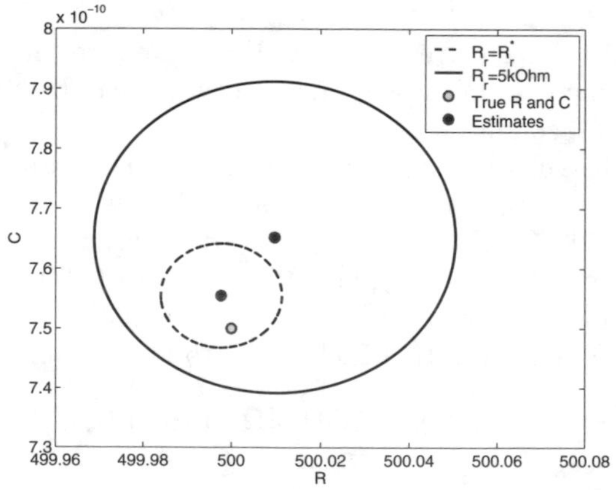


**Fig. 5.4.** Confidence region (5.25) of $R$ and $C$ for $R_r = 5[\mathrm{k}\Omega]$ (larger) and $R_r = R_r^*$, respectively

It can be observed from Fig. 5.4 and the inequality (5.25) that the hypothesis $\mathcal{H}_0$ is rejected when $R_r = R_r^*$, which means that a fault occurs. On the contrary, the hypothesis $\mathcal{H}_0$ is accepted when $R_r = 5[\mathrm{k}\Omega]$, which means that there is no fault. These results clearly indicate that the application of D-optimum experimental conditions increases fault sensitivity, i.e., it makes the proposed fault diagnosis scheme more reliable.

### Accuracy analysis

The main objective of this section is to estimate the measurement accuracy provided by the approach considered. For that purpose, a set of different impedances were selected (similar to that of [5]). Each measurement was repeated 50 times, and then the mean measured values $\bar{R}$ and $\bar{C}$ were calculated and for each of them a coefficient of variation $\bar{\sigma}$ was computed:

$$\bar{\sigma} = \frac{\sigma_R}{\bar{R}}100[\%], \quad \text{or} \quad \bar{\sigma} = \frac{\sigma_C}{\bar{C}}100[\%], \tag{5.56}$$

where $\sigma_C$ (or $\sigma_R$) stands for the standard deviation of 50 measurements. Table 5.1 shows the achieved results. From this it is clear that the proposed approach provides a high measurement accuracy. It should also be pointed out that these measurements were achieved for $n_t = 4000$, which implies that the measurement time was 1[s]. Figure 5.5 shows the evolution of the relative errors

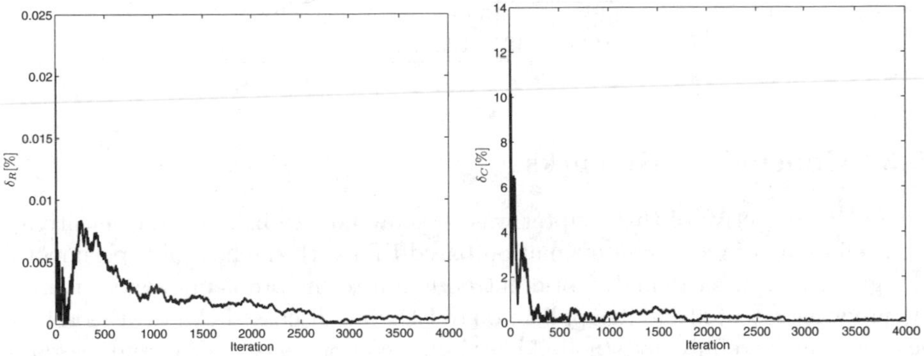


**Fig. 5.5.** Relative errors in the consecutive iterations of the proposed algorithm

(5.54) (for $R = 500[\Omega]$, $C = 0.75[\mathrm{nF}]$, and $R_r = R_r^*$) in the consecutive iterations of the proposed algorithm. From these results it is clear that relatively high measurement accuracies can be achieved after a few hundred iterations only. This corresponds to the measurement time less than 0.25[s].

**Table 5.1.** Simulation results

| | True value | Mean measured value | $\bar{\sigma}[\%]$ |
|---|---|---|---|
| C | 0.75[nF] | 0.7502[nF] | 0.41 |
| R | 500[$\Omega$] | 499.9997[$\Omega$] | $9.2252 \cdot 10^{-4}$ |
| C | 150[nF] | 150[nF] | 0.0018 |
| R | 570[$\Omega$] | 569.999[$\Omega$] | $9.3179 \cdot 10^{-4}$ |
| C | 320[nF] | 320[nF] | $4.6464 \cdot 10^{-4}$ |
| R | 48[k$\Omega$] | 4.7999[k$\Omega$] | 0.0425 |
| C | 1[nF] | 0.9998[nF] | 0.1441 |
| R | 1[k$\Omega$] | 999.99[$\Omega$] | 0.0010 |
| C | 50[nF] | 50[nF] | 0.0028 |
| R | 1.1[k$\Omega$] | 1.1[k$\Omega$] | $8.7803 \cdot 10^{-4}$ |
| C | 160[nF] | 160[nF] | $5.1671 \cdot 10^{-4}$ |
| R | 97[k$\Omega$] | 96.986[k$\Omega$] | 0.0391 |
| C | 840[pF] | 840[pF] | 0.0313 |
| R | 5[k$\Omega$] | 5[k$\Omega$] | $8.6657 \cdot 10^{-4}$ |
| C | 15[nF] | 15[nF] | 0.0018 |
| R | 5.7[k$\Omega$] | 5.7[k$\Omega$] | $9.1961 \cdot 10^{-4}$ |
| C | 32[nF] | 32[nF] | $4.5223 \cdot 10^{-4}$ |
| R | 296[k$\Omega$] | 296[k$\Omega$] | 0.0301 |
| C | 540[pF] | 540[pF] | 0.0220 |
| R | 10[k$\Omega$] | 10[k$\Omega$] | $8.8802 \cdot 10^{-4}$ |
| C | 13[nF] | 13[nF] | $8.8790 \cdot 10^{-4}$ |
| R | 17[k$\Omega$] | 17[k$\Omega$] | 0.0012 |
| C | 16[nF] | 16[nF] | $5.3046 \cdot 10^{-4}$ |
| R | 495[k$\Omega$] | 495.01[k$\Omega$] | 0.0245 |

## 5.3 Concluding Remarks

The main objective of this chapter was to show how to increase the sensitivity and reliability of parameter estimation-based FDI with the help of experimental design theory. In particular, a brief introduction to the problem of experimental design was presented and its application to FDI was discussed. All the theoretical aspects were carefully investigated based on a comprehensive case study regarding impedance measurement. Starting from a general problem, it was shown how to construct the so-called virtual bridge. Subsequently, it was shown that the impedance measurement problem can be formulated as a non-linear parameter estimation task. This task can then be reduced to a linear parameter estimation problem without any linearisation. It was also shown how to obtain parameter confidence region that can be used for FDI. Based on the achieved results, it was shown how to determine optimum experimental conditions. The experimental results clearly show the benefits that can be gained while using the experimental design both from the parameter estimation and fault diagnosis viewpoints. The final conclusion that can be drawn is that works dealing with experimental design for FDI are relatively scarce. Indeed, the optimisation of the data acquisition process, which, undoubtedly, increases the reliability of fault diagnosis, is most often neglected. This is especially the case for non-linear dynamic systems.

# Part III

# Soft Computing Strategies

# 6. Evolutionary Algorithms

As was mentioned in Section 3.2, evolutionary algorithms are powerful optimisation tools which can be applied to various challenging problems for which the classic optimisation techniques cannot be employed or do not give satisfactory results. Section 3.2.2 presents a large spectrum of FDI problems that can be solved with EAs. This bibliographic review clearly shows that there are two general approaches that deserve particular attention, i.e., the integration of analytical and soft computing FDI techniques and robust soft computing-based FDI. In this chapter, the attention is drawn to the first aspect.

As was indicated in Chapter 2, observers are immensely popular as residual generators for fault detection (and, consequently, for fault isolation) of both linear and non-linear dynamic systems. Their popularity lies in the fact that they can also be employed for control purposes. There are, of course, many different observers which can be applied to non-linear and especially non-linear deterministic systems, and the best known of them were briefly reviewed in Section 2.2.3. Logically, the number of "real world" applications of observers (not only simulated examples) should proliferate, yet this is not the case. It seems that there are two main reasons why strong formal methods are not accepted in engineering practice. First, the application of observers is limited by the need for non-linear state-space models of the system being considered, which is usually a serious problem in complex industrial systems. This explains why most of the examples considered in the literature are devoted to simulated or laboratory systems, e.g., the celebrated three- (two- or even four-) tank system, an inverted pendulum, a traveling crane, etc. Thus, one possible application of evolutionary algorithms is to use them for model design. Such an interesting approach [183, 187] was briefly described in Section 3.2.2. Another reason is that the design complexity of most observers for non-linear systems does not encourage engineers to apply them in practice. To tackle this problem, Chapter 4 provides three unknown input observer structures with relatively simple design procedures. The first two observers are based on the general idea of an extended unknown input observer [187]. Thus, the main design objective is to improve their convergence rate. Sections 4.1 and 4.2 show analytical solutions to this challenging task.

M. Witczak: Model. and Estim. Strat. for Fault Diagn. of Non-Linear Syst. LNCIS 354, pp. 105–132, 2007.
springerlink.com 

However, as was indicated in Section 3.2.2 it is possible to increase the convergence rate further with the help of evolutionary algorithms.

Taking into account the above discussion, the chapter is organised as follows: Section 6.1 presents a genetic programming approach that can be used for designing models of non-linear systems. In Section 6.2, genetic programming is used to improve the convergence rate of the EUIO described in Section 4.1. Similarly, Section 6.3 presents the application of the ESSS algorithm [63] and a stochastic robustness technique for improving the convergence rate of the EUIO described in Section 4.2. Finally, Section 6.4 presents selected experimental results obtained with the above-described techniques.

## 6.1 Model Design with Genetic Programming

Generally, the model determination process can be realised as follows:

*Step 0:* Select a set of possible model structures $\mathbb{M}$.
*Step 1:* Estimate the parameters of each of the models $M_i,\ i = 1, \ldots, n_m$.
*Step 2:* Select the model which is best suited in terms of the selected criterion.
*Step 3:* If the selected model does not satisfy the prespecified requirements, then go to *Step 0.*

The above four-step procedure is usually very complex and requires advanced experience regarding the system being modelled. The subsequent part of this section shows how to automate the above procedure with genetic programming [183, 187].

The characterisation of a set of possible candidate models $\mathbb{M}$ from which the system model will be obtained constitutes an important preliminary task in any system identification procedure. Knowing that the system exhibits a non-linear characteristic, a choice of a non-linear model set must be made. Let a non-linear input-output MIMO model have the following form:

$$\begin{aligned} y_{M,i,k} =& g_i\big(y_{M,1,k-1}, \ldots, y_{M,1,k-n_{1,y}}, \ldots, y_{M,m,k-1}, \ldots, y_{M,m,k-n_{m,y}}, \\ & u_{1,k-1}, \ldots, u_{1,k-n_{1,u}}, \ldots, u_{r,k-1}, \ldots, u_{r,k-n_{r,u}}, \boldsymbol{p}_i\big), \\ & i = 1, \ldots, m. \end{aligned} \tag{6.1}$$

Thus the system output is given by

$$\boldsymbol{y}_k = \boldsymbol{y}_{M,k} + \boldsymbol{\varepsilon}_k, \tag{6.2}$$

where $\boldsymbol{\varepsilon}_k$ consists of a structural deterministic error caused by the model-reality mismatch and the stochastic error caused by the measurement noise $\boldsymbol{v}_k$. The problem is to determine the unknown function $\boldsymbol{g}(\cdot) = (g_1(\cdot), \ldots g_m(\cdot))$ and to estimate the corresponding parameter vector $\boldsymbol{p} = [\boldsymbol{p}_1^T, \ldots, \boldsymbol{p}_m^T]^T$.

One possible solution to this problem is the GP approach. As has already been mentioned, the main ingredient underlying the GP algorithm is a tree. In order to adapt GP to system identification, it is necessary to represent the model (6.1) either as a tree or as a set of trees. Indeed, as is shown in Fig. 6.1,

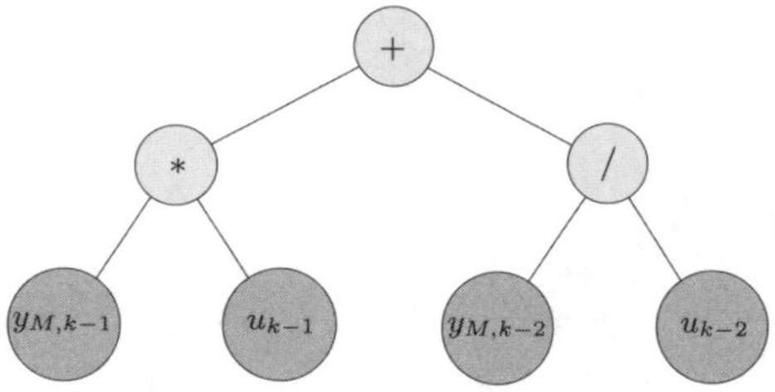


**Fig. 6.1.** Exemplary GP tree representing the model $y_{M,k} = y_{M,k-1}u_{k-1} + y_{M,k-2}/u_{k-2}$

the MISO model can be easily put in the form of a tree, and hence to build the MIMO model (6.1) it is necessary to use $m$ trees. In such a tree (see Fig. 6.1), two sets can be distinguished, namely, the terminal set $\mathbb{T}$ and the function set $\mathbb{F}$ (e.g., $\mathbb{T} = \{u_{k-1}, u_{k-2}, y_{M,k-1}, y_{M,k-2}\}$, $\mathbb{F} = \{+, *, /\}$). The language of the trees in GP is formed by a user-defined function $\mathbb{F}$ set and a terminal $\mathbb{T}$ set, which form the nodes of the trees. The functions should be chosen so as to be a priori useful in solving the problem, i.e., any knowledge concerning the system under consideration should be included in the function set. This function set is very important and should be universal enough to be capable of representing a wide range of non-linear systems. The terminals are usually variables or constants. Thus, the searching space consists of all the possible compositions that can be recursively formed from the elements of $\mathbb{F}$ and $\mathbb{T}$. The selection of variables does not cause any problems, but the handling of numerical parameters (constants) seems very difficult. Even though there are no constant numerical values in the terminal set $\mathbb{T}$, they can be implicitly generated, e.g., the number 0.5 can be expressed as $x/(x + x)$. Unfortunately, such an approach leads to an increase in both the computational burden and evolution time. Another way is to introduce a number of random constants into the terminal set, but this is also an inefficient approach. An alternative way of handling numerical parameters which seems to be more suitable is called *node gains* [47]. A node gain is a numerical parameter associated with the node whose output it multiplies (see Fig. 6.2). Although this technique is straightforward, it leads to an excessive number of parameters, i.e., there are parameters which are not identifiable. Thus, it is necessary to develop a mechanism which prevents such situations from happening. First, let us define the function set $\mathbb{F} = \{+, *, /, \xi_1(\cdot), \ldots, \xi_l(\cdot)\}$, where $\xi_k(\cdot)$ is a non-linear univariate function. To tackle the parameter reduction problem, several simple rules can be established [183, 187]:

- $*, /$: A node of type either $*$ or $/$ always has parameters set to unity on the side of its successors. If a node of the above type is a root node of a tree, then the parameter associated with it should be estimated.
- $+$: A parameter associated with a node of type $+$ is always equal to unity. If its successor is not of type $+$, then the parameter of the successor should be estimated.

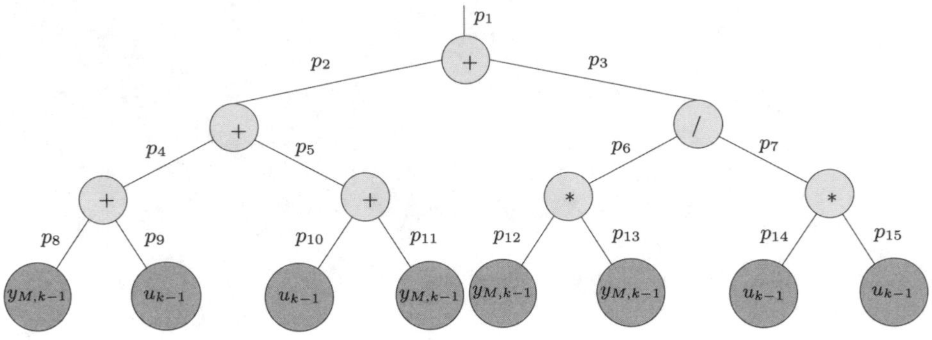


**Fig. 6.2.** Exemplary parameterised tree

$\xi$: If a successor of a node of type $\xi$ is a leaf of a tree or is of type $*$ or $/$, then the parameter of the successor should be estimated. If a node of type $\xi$ is a root of a tree, then the associated parameter should be estimated.

As an example, consider the tree shown in Fig. 6.2. Following the above rules, the resulting parameter vector has only five elements $\boldsymbol{p} = (p_3, p_8, p_9, p_{10}, p_{11})$, and the resulting model is $y_{M,k} = (p_{11}+p_8)y_{M,k-1}+(p_{10}+p_9)u_{k-1}+p_3 y_{M,k-1}^2/u_{k-1}^2$. It is obvious that although these rules are not optimal in the sense of parameter identifiability, their application significantly reduces the dimension of the parameter vector, thus making the parameter estimation process much easier. Moreover, the introduction of parameterised trees reduces the terminal set to variables only, i.e., constants are no longer necessary, and hence the terminal set is given by

$$\begin{aligned}\mathbb{T} = \{&y_{M,1,k-1}, \dots, y_{M,1,k-n_{1,y}}, \dots, y_{M,m,k-1}, \dots, y_{M,m,k-n_{m,y}},\\ &u_{1,k-1}, \dots, u_{1,k-n_{1,u}}, \dots, u_{r,k-1}, \dots, u_{r,k-n_{r,u}}\}.\end{aligned} \tag{6.3}$$

The remaining problem is to select appropriate lags in the input and output signals of the model. Assuming that $n_y^{\max}$ i $n_u^{\max}$ are maximum lags in the output and input signals, the problem boils down to checking $n_y^{\max} \times n_u^{\max}$ possible configurations, which is an extremely time-consuming process. With a slight loss of generality, it is possible to assume that each $n_y = n_u = n$. Thus the problem reduces to finding, throughout experiments, such $n$ for which the model is the best replica of the system. It should also be pointed out that the true parameter vector $\boldsymbol{p}$ is unknown and hence it should replaced by its current estimate $\hat{\boldsymbol{p}}$. Consequently, instead of using $\boldsymbol{y}_M$ in the terminal set $\mathbb{T}$ the output estimate $\hat{\boldsymbol{y}}$ should be employed, which can easily be calculated with (6.1) and $\hat{\boldsymbol{p}}$.

### 6.1.1 Model Structure Determination Using GP

If the terminal and function sets are given, populations of GP individuals (trees) can be generated, i.e., the set $\mathbb{M}$ of possible model structures is created. An outline of the GP algorithm is shown in Tab. 6.1. The algorithm works on a set

**Table 6.1.** Outline of the GP algorithm

*I. Initiation*

A. *Random generation* $\mathbb{P}(0) = \{P_i(0) \mid i = 1, \ldots, n_{\text{pop}}\}$.

B. *Fitness calculation* $\Phi(\mathbb{P}(0)) = \{\Phi(P_i(0)) \mid i = 1, \ldots, n_{\text{pop}}\}$.

C. $t = 1$.

*II. While* $(\iota(\mathbb{P}(t)) = \text{true})$ *do*

A. *Selection* $\mathbb{P}'(t) = \{P_i'(t) = s_{n_s}(P_i(t)) \mid i = 1, \ldots, n_{\text{pop}}\}$.

B. *Crossover* $\mathbb{P}''(t) = \{P_i''(t) = r_{P_{cross}}(P_i'(t)) \mid i = 1, \ldots, n_{\text{pop}}\}$.

C. *Mutation* $\mathbb{P}'''(t) = \{P_i'''(t) = m_{P_{mut}}(P_i''(t)) \mid i = 1, \ldots, n_{\text{pop}}\}$.

D. *Fitness calculation* $\Phi(\mathbb{P}'''(t)) = \{\Phi(P_i'''(t)) \mid i = 1, \ldots, n_{\text{pop}}\}$.

E. *New generation* $\mathbb{P}(t+1) = \{P_i(t+1) = P_i'''(t) \mid i = 1, \ldots, n_{\text{pop}}\}$.

F. $t = t + 1$.

of populations $\mathbb{P} = \{P_i \mid i = 1, \ldots, n_{\text{pop}}\}$, and the number of populations $n_{\text{pop}}$ depends on the application, e.g., in the case of the model (6.1), the number of populations is equal to the dimension $m$ of the output vector $\boldsymbol{y}_k$, i.e., $n_{\text{pop}} = m$. Each of the above populations $P_i = \{b_{i,j} \mid j = 1, \ldots, n_m\}$ is composed of a set of $n_m$ trees $b_{i,j}$. Since the number of populations is given, the GP algorithm can be started (*initiation*) by randomly generating individuals, i.e., $n_m$ individuals are created in each population whose trees are of a desired depth $n_d$. The tree generating process can be performed in several different ways, resulting in trees of different shapes. The basic approaches are the "full" and "grow" methods [103]. The "full" method generates trees for which the length of every non-backtracking path from the root to an endpoint is equal to the prespecified depth $n_d$. The "grow" method generates trees of various shapes. The length of a path between the root and an endpoint is not greater than the prespecified depth $n_d$. Because of the fact that, in general, the shape of the true solution is unknown, it seems desirable to combine both of the above methods. Such a combination is called *ramped half-and-half*. Moreover, it is assumed that the parameters $\boldsymbol{p} = [\boldsymbol{p}_1^T, \ldots, \boldsymbol{p}_m^T]^T$ of each tree are initially set to unity (although it is possible to set the parameters randomly). In the first step (*fitness calculation*), the estimation of the parameter vector $\boldsymbol{p}$ of each individual is performed, according to some predefined criterion, e.g., the Akaike Information Criterion (AIC) [170], which is related to the fitness function $\Phi(\cdot)$. In the case of parameter estimation, many algorithms can be employed; more precisely, as GP models are usually non-linear in their parameters, the choice reduces to one of non-linear optimisation techniques.

Unfortunately, because models are randomly generated, they can contain linearly dependent parameters (even after the application of parameter reduction rules) and parameters which have very little influence on the model output. In many cases, this may lead to a very pure performance of gradient-based algorithms. Owing to the above-mentioned problems, the spectrum of possible non-linear optimisation techniques reduces to gradient-free techniques, which usually require a large number of cost evaluations. On the other hand, the application of stochastic gradient-free algorithms, apart from the simplicity of the approach, decreases the chance to get stuck in a local optimum, and hence it may give more suitable parameter estimates. Based on numerous computer experiments, it has been found that the extremely simple Adaptive Random Search (ARS) algorithm [170] is especially well suited for that purpose. The routine chooses the initial parameter vector $\boldsymbol{p}^0$, e.g., $\boldsymbol{p}^0 = \mathbf{1}$. After $q$ iterations, given the current best estimate $\boldsymbol{p}^q$, a random displacement vector $\boldsymbol{\Delta p}$ is generated and the trial point

$$\boldsymbol{p}^* = \boldsymbol{p}^q + \boldsymbol{\Delta p} \tag{6.4}$$

is checked, with $\boldsymbol{\Delta p}$ following the normal distribution with zero-mean and covariance

$$\boldsymbol{\Sigma} = \mathrm{diag}[\sigma_1, \ldots, \sigma_{\dim \boldsymbol{p}}]. \tag{6.5}$$

If $\Phi(b_{i,j}(\boldsymbol{p}^*)) > \Phi(b_{i,j}(\boldsymbol{p}^q))$, then $\boldsymbol{p}^*$ is rejected and, consequently, $\boldsymbol{p}^{q+1} = \boldsymbol{p}^q$ is set; otherwise, $\boldsymbol{p}^{q+1} = \boldsymbol{p}^*$ The adaptive strategy consists in repeatedly alternating two phases. During the first one (*variance selection*), $\boldsymbol{\Sigma}$ is selected from the sequence ${}^1\sigma, \ldots, {}^5\sigma$, where ${}^1\sigma$ is set by the user in such a way as to allow an easy exploration of the parameter space, and

$${}^i\sigma = {}^{i-1}\sigma/10, \quad i = 2, \ldots, 5. \tag{6.6}$$

In order to allow a comparison to be drawn, all the possible ${}^i\sigma$s are used $100/i$ times, starting from the same initial value of $\boldsymbol{p}$. The largest ${}^i\sigma$s, designed to escape the local minimum, are therefore used more often than the smaller ones. During the second (exploration) phase, the most successful ${}^i\sigma$ is used to perform 100 random trials starting from the best $\boldsymbol{p}$ obtained so far.

In the next step, the fitness of each model is obtained and the best-suited model is selected. If the selected model satisfies the prespecified requirements, then the algorithm is stopped. In the second step, the *selection* process is applied to create a new intermediate population of parent individuals. For that purpose, various approaches can be employed, e.g., proportional selection, rank selection, tournament selection [103, 115]. The selection method used in the present book is tournament selection, and it works as follows: select randomly $n_s$ models, i.e., trees which represent the models, and copy the best of them into the intermediate set of models (intermediate populations). The above procedure is repeated $n_m$ times.

The individuals for the new population (the next generation) are produced through the application of *crossover* and *mutation*. To apply *crossover* $r_{P_{\text{cross}}}$,

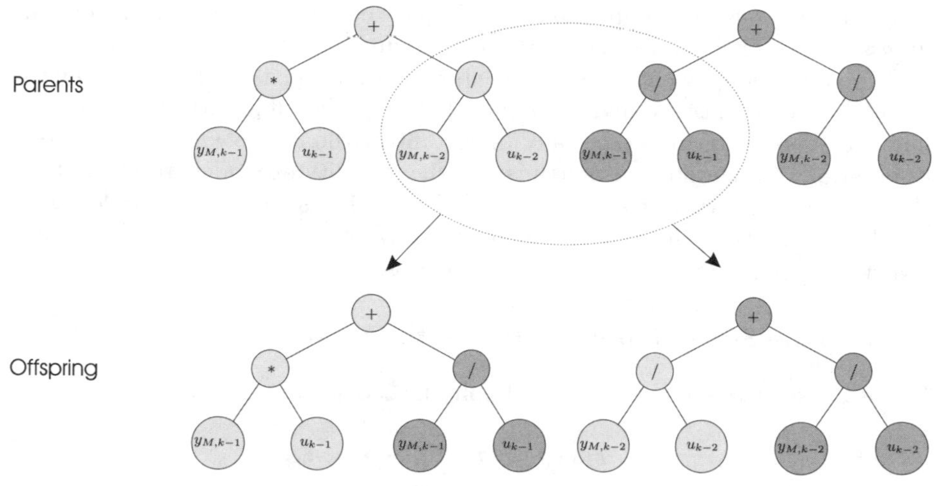


**Fig. 6.3.** Exemplary crossover operation

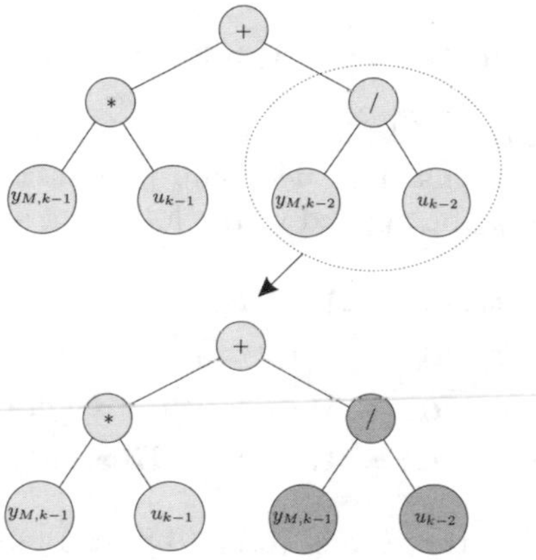

**Fig. 6.4.** Exemplary mutation operation

random couples of individuals which have the same position in each population are formed. Then, with a probability $P_{\text{cross}}$, each couple undergoes crossover, i.e., a random crossover point (a node) is selected and then the corresponding sub-trees are exchanged (Fig. 6.3). *Mutation* $m_{P_{mut}}$ (Fig. 6.4) is implemented in such a way that for each entry of each individual a sub-tree at a selected point is removed with probability $P_{\text{mut}}$ and replaced with a randomly generated tree. The parameter vectors of individuals which have been modified by means of either crossover or mutation are set to unity (although a different choice is possible), while the parameters of the remaining individuals are unchanged. The GP algorithm is repeated until the best-suited model satisfies the prespecified requirements $\iota(\mathbb{P}(t))$, or until the number of maximum admissible iterations has been exceeded. It should also be pointed out that the simulation programme must ensure robustness to unstable models. This can easily be attained when the fitness function is bounded by a certain maximum admissible value. This means that each individual which exceeds the above bound is penalised by stopping the calculation of its fitness, and then $\Phi(b_{i,j})$ is set to a sufficiently large positive number. This problem is especially important in the case of input-output representation of the system. Unfortunately, the stability of models resulting from this approach is very difficult to prove. However, this is a common problem with non-linear input-output dynamic models. To overcome it, an alternative model structure is presented in the subsequent section.

### 6.1.2 State-Space Representation of the System

Let us consider the following class of non-linear discrete-time systems:

$$\boldsymbol{x}_{k+1} = \boldsymbol{g}\left(\boldsymbol{x}_k, \boldsymbol{u}_k\right) + \boldsymbol{w}_k, \tag{6.7}$$

$$\boldsymbol{y}_{k+1} = \boldsymbol{C}\boldsymbol{x}_{k+1} + \boldsymbol{v}_k. \tag{6.8}$$

Assume that the function $\boldsymbol{g}\left(\cdot\right)$ has the form

$$\boldsymbol{g}\left(\boldsymbol{x}_k, \boldsymbol{u}_k\right) = \boldsymbol{A}(\boldsymbol{x}_k)\boldsymbol{x}_k + \boldsymbol{h}(\boldsymbol{u}_k). \tag{6.9}$$

The choice of the structure (6.9) is caused by the fact that the resulting model is to be used in FDI systems. The algorithm presented below though can also, with minor modifications, be applied to the following structures of $\boldsymbol{g}\left(\cdot\right)$:

$$\boldsymbol{g}\left(\boldsymbol{x}_k, \boldsymbol{u}_k\right) = \boldsymbol{A}(\boldsymbol{x}_k, \boldsymbol{u}_k)\boldsymbol{x}_k, \tag{6.10}$$

$$\boldsymbol{g}\left(\boldsymbol{x}_k, \boldsymbol{u}_k\right) = \boldsymbol{A}(\boldsymbol{x}_k, \boldsymbol{u}_k)\boldsymbol{x}_k + \boldsymbol{h}(\boldsymbol{u}_k), \tag{6.11}$$

$$\boldsymbol{g}\left(\boldsymbol{x}_k, \boldsymbol{u}_k\right) = \boldsymbol{A}(\boldsymbol{x}_k, \boldsymbol{u}_k)\boldsymbol{x}_k + \boldsymbol{B}(\boldsymbol{x}_k)\boldsymbol{u}_k, \tag{6.12}$$

$$\boldsymbol{g}\left(\boldsymbol{x}_k, \boldsymbol{u}_k\right) = \boldsymbol{A}(\boldsymbol{x}_k)\boldsymbol{x}_k + \boldsymbol{B}(\boldsymbol{x}_k)\boldsymbol{u}_k. \tag{6.13}$$

The state-space model of the system (6.7)–(6.8) can be expressed as

$$\boldsymbol{x}_{k+1} = \boldsymbol{A}(\boldsymbol{x}_k)\boldsymbol{x}_k + \boldsymbol{h}(\boldsymbol{u}_k), \tag{6.14}$$

$$\boldsymbol{y}_{M,k+1} = \boldsymbol{C}\boldsymbol{x}_{k+1}. \tag{6.15}$$

The problem is to determine the matrices $\boldsymbol{A}(\cdot)$, $\boldsymbol{C}$ and the vector $\boldsymbol{h}(\cdot)$, given the set of input-output measurements $\{(\boldsymbol{u}_k, \boldsymbol{y}_k)\}_{k=0}^{n_t-1}$. Moreover, it is assumed that the true state vector $\boldsymbol{x}_k$ is, in particular, unknown. This means that its estimate $\hat{\boldsymbol{x}}_k$ (obtained for a given model) should be used instead. Without loss of generality, it is possible to assume that

$$\boldsymbol{A}(\boldsymbol{x}_k) = \operatorname{diag}[a_{1,1}(\boldsymbol{x}_k), a_{2,2}(\boldsymbol{x}_k), \ldots, a_{n,n}(\boldsymbol{x}_k)]. \tag{6.16}$$

Thus, the problem reduces to identifying the non-linear functions $a_{i,i}(\boldsymbol{x}_k)$, $h_i(\boldsymbol{u}_k)$, $i = 1, \ldots, n$, and the matrix $\boldsymbol{C}$. Now it is possible to establish the conditions under which the model (6.14)–(6.15) is globally asymptotically stable.

**Theorem 6.1.** *If, for* $\boldsymbol{h}(\boldsymbol{u}_k) = \boldsymbol{0}$,

$$\forall k \geq 0, \quad \forall \boldsymbol{x}_k \in \mathbb{R}^n, \quad \max_{i=1,\ldots,n} |a_{i,i}(\boldsymbol{x}_k)| < 1, \tag{6.17}$$

*then the model (6.14)–(6.15) is globally asymptotically stable, i.e.,* $\boldsymbol{x}_k$ *converges to the equilibrium point* $\boldsymbol{x}^*$ *for any* $\boldsymbol{x}_0$.

*Proof.* Since the matrix $\boldsymbol{A}(\boldsymbol{x}_k)$ is a diagonal one,

$$\|\boldsymbol{A}(\boldsymbol{x}_k)\| = \max_{i=1,\ldots,n} |\lambda_i(\boldsymbol{A}(\boldsymbol{x}_k))| = \max_{i=1,\ldots,n} |a_{i,i}(\boldsymbol{x}_k)|, \tag{6.18}$$

where the norm $\|\boldsymbol{A}(\cdot)\|$ may have one of the following forms:

$$\|\boldsymbol{A}(\cdot)\|_2 = \sqrt{\lambda_{\max}(\boldsymbol{A}(\cdot)^T\boldsymbol{A}(\cdot))}, \tag{6.19}$$

$$\|\boldsymbol{A}(\cdot)\|_1 = \max_{1\leq i\leq n} \sum_{j=1}^{n} |a_{i,j}(\cdot)|, \tag{6.20}$$

$$\|\boldsymbol{A}(\cdot)\|_\infty = \max_{1\leq j\leq n} \sum_{i=1}^{n} |a_{i,j}(\cdot)|. \tag{6.21}$$

Finally, using [21, Proof of Theorem 1] yields the condition (6.17).

Since the stability conditions are established, it is possible to give a general framework for the identification of (6.14)–(6.15). Since $a_{i,i}(\boldsymbol{x}_k)$, $h_i(\boldsymbol{u}_k)$, $i = 1,\ldots,n$ are assumed to be non-linear (in general) functions, it is necessary to use $n$ populations to represent $a_{i,i}(\boldsymbol{x}_k)$, $i = 1,\ldots,n$, and another $n$ populations to represent $h_i(\boldsymbol{u}_k)$, $i = 1,\ldots,n$. Thus the number of populations is $n_{\text{pop}} = 2n$. The terminal sets for these two kinds of populations are different, i.e., the first terminal set is defined as $\mathbb{T}_A = \{x_{1,k},\ldots,x_{n,k}\}$, and the second one as $\mathbb{T}_h = \{u_{1,k},\ldots,u_{r,k}\}$. The parameter vector $\boldsymbol{p}$ consists of the parameters of both $a_{i,i}(\boldsymbol{x}_k)$ and $h_i(\boldsymbol{u}_k)$. Unfortunately, the estimation of $\boldsymbol{p}$ is not as simple as in the input-output representation case. This means that checking the trial point in the ARS algorithm (see Section 6.1.1) involves the computation of $\boldsymbol{C}$, which is necessary to obtain the output error $\boldsymbol{\varepsilon}_k$ and, consequently, the value of the fitness function. To tackle this problem, for each trial point $\boldsymbol{p}$ it is necessary to first set an initial state estimate $\hat{\boldsymbol{x}}_0$, and then to obtain the state estimate $\hat{\boldsymbol{x}}_k$, $k = 1,\ldots,n_t-1$. Knowing the state estimate and using the least-square method, it is possible to obtain $\boldsymbol{C}$ by solving the following equation:

$$\boldsymbol{C}\sum_{k=0}^{n_t-1} \hat{\boldsymbol{x}}_k\hat{\boldsymbol{x}}_k^T = \sum_{k=0}^{n_t-1} \boldsymbol{y}_k\hat{\boldsymbol{x}}_k^T, \tag{6.22}$$

or, equivalently, by using

$$\boldsymbol{C} = \sum_{k=0}^{n_t-1} \boldsymbol{y}_k\hat{\boldsymbol{x}}_k^T \left[\sum_{k=0}^{n_t-1} \hat{\boldsymbol{x}}_k\hat{\boldsymbol{x}}_k^T\right]^{-1}. \tag{6.23}$$

Since the identification procedure of (6.14)–(6.15) is given, it is possible to establish the structure of $\boldsymbol{A}(\cdot)$, which guarantees that the condition of Theorem 6.1 is always satisfied, i.e., $\max_{i=1,\dots,n} |a_{i,i}(\boldsymbol{x}_k)| < 1$. This can easily be achieved with the following structure of $a_{i,i}(\boldsymbol{x}_k)$:

$$a_{i,i}(\boldsymbol{x}_k) = \tanh(s_{i,i}(\boldsymbol{x}_k)), \quad i = 1, \dots, n, \tag{6.24}$$

where $\tanh(\cdot)$ is a hyperbolic tangent function, and $s_{i,i}(\boldsymbol{x}_k)$ is a function represented by the GP tree. It should also be pointed out that the order $n$ of the model is in general unknown, and hence it should be determined throughout experiments.

## 6.2 Robustifying the EUIO with Genetic Programming

Although the approach to improving the convergence rate of the EUIO presented in Section 4.1 is very simple, it is possible to increase the convergence rate further. Indeed, the structure (4.59) is probably not the best solution for setting $\boldsymbol{Q}_k$ and $\boldsymbol{R}_k$ in all potential applications. A more general approach is to set the instrumental matrices as follows:

$$\boldsymbol{Q}_{k-1} = q^2(\boldsymbol{\varepsilon}_{k-1})\boldsymbol{I} + \delta_1 \boldsymbol{I}, \quad \boldsymbol{R}_k = r^2(\boldsymbol{\varepsilon}_k)\boldsymbol{I} + \delta_2 \boldsymbol{I}, \tag{6.25}$$

while $q(\boldsymbol{\varepsilon}_{k-1})$ and $r(\boldsymbol{\varepsilon}_k)$ are non-linear functions of the output error $\boldsymbol{\varepsilon}_k$ (the squares are used to ensure the positive definiteness of $\boldsymbol{Q}_{k-1}$ and $\boldsymbol{R}_k$). Thus, the problem reduces to identifying the above functions. To tackle it, genetic programming can be employed. The unknown functions $q(\boldsymbol{\varepsilon}_{k-1})$ and $r(\boldsymbol{\varepsilon}_k)$ can be expressed as trees, as shown in Fig. 6.5. Thus, in the case of $q(\cdot)$ and $r(\cdot)$, the terminal sets are $\mathbb{T}_Q = \{\varepsilon_{1,k-1}, \dots, \varepsilon_{m,k-1}\}$ and $\mathbb{T}_R = \{\varepsilon_{1,k}, \dots, \varepsilon_{m,k}\}$, respectively. In both cases, the function set can be defined as $\mathbb{F} = \{+, *, /, \xi_1(\cdot), \dots, \xi_l(\cdot)\}$, where $\xi_k(\cdot)$ is a non-linear univariate function and, consequently, the number of populations is $n_{\text{pop}} = 2$. Since the terminal and function sets are given, the approach described in Section 6.1 can easily be adapted for the identification

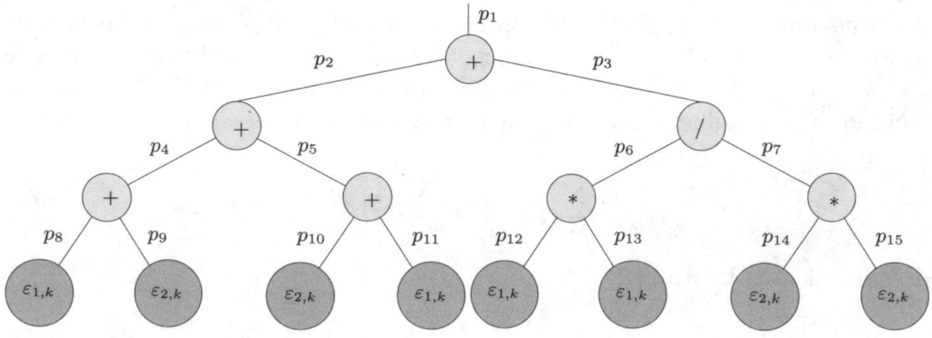


**Fig. 6.5.** Exemplary tree representing $r(\boldsymbol{\varepsilon}_k)$

purpose of $q(\cdot)$ and $r(\cdot)$. First, let us define the identification criterion constituting a necessary ingredient of the $\boldsymbol{Q}_{k-1}$ and $\boldsymbol{R}_k$ selection process.

Since the instrumental matrices should be chosen so as to satisfy (4.35)–(4.36), the selection of $\boldsymbol{Q}_{k-1}$ and $\boldsymbol{R}_k$ can be performed according to

$$\left(\boldsymbol{Q}_{k-1}, \boldsymbol{R}_k\right) = \arg \max_{q(\boldsymbol{\varepsilon}_{k-1}), r(\boldsymbol{\varepsilon}_k)} j_{\text{obs},1}\left(q(\boldsymbol{\varepsilon}_{k-1}), r(\boldsymbol{\varepsilon}_k)\right), \tag{6.26}$$

where

$$j_{\text{obs},1}\left(q(\boldsymbol{\varepsilon}_{k-1}), r(\boldsymbol{\varepsilon}_k)\right) = \sum_{k=0}^{n_t-1} \text{trace}\boldsymbol{P}_k. \tag{6.27}$$

On the other hand, owing to FDI requirements, it is clear that the output error should be near zero in the fault-free mode. In this case, one can define another identification criterion:

$$\left(\boldsymbol{Q}_{k-1}, \boldsymbol{R}_k\right) = \arg \min_{q(\boldsymbol{\varepsilon}_{k-1}), r(\boldsymbol{\varepsilon}_k)} j_{\text{obs},2}\left(q(\boldsymbol{\varepsilon}_{k-1}), r(\boldsymbol{\varepsilon}_k)\right), \tag{6.28}$$

and

$$j_{\text{obs},2}\left(q(\boldsymbol{\varepsilon}_{k-1}), r(\boldsymbol{\varepsilon}_k)\right) = \sum_{k=0}^{n_t-1} \boldsymbol{\varepsilon}_k^T \boldsymbol{\varepsilon}_k. \tag{6.29}$$

Therefore, in order to join (6.26) and (6.28), the following identification criterion is employed:

$$\left(\boldsymbol{Q}_{k-1}, \boldsymbol{R}_k\right) = \arg \min_{q(\boldsymbol{\varepsilon}_{k-1}), r(\boldsymbol{\varepsilon}_k)} j_{\text{obs},3}\left(q(\boldsymbol{\varepsilon}_{k-1}), r(\boldsymbol{\varepsilon}_k)\right), \tag{6.30}$$

where

$$j_{\text{obs},3}\left(q(\boldsymbol{\varepsilon}_{k-1}), r(\boldsymbol{\varepsilon}_k)\right) = \frac{j_{\text{obs},2}\left(q(\boldsymbol{\varepsilon}_{k-1}), r(\boldsymbol{\varepsilon}_k)\right)}{j_{\text{obs},1}\left(q(\boldsymbol{\varepsilon}_{k-1}), r(\boldsymbol{\varepsilon}_k)\right)}. \tag{6.31}$$

Since the identification criterion is established, it is straightforward to use the GP algorithm detailed in Section 6.1.

## 6.3 Robustifying the EUIO with the ESSS Algorithm

Although the settings of $\boldsymbol{Q}_k$ and $\boldsymbol{R}_k$ determined by (4.127) and (4.131) are very straightforward and easy to implement, it is possible to increase the convergence rate further. As was shown in Section 6.2, the form of the instrumental matrices $\boldsymbol{Q}_k$ and $\boldsymbol{R}_k$ can be determined with the genetic programming approach.

A completely different approach was proposed in [72]. In particular, the authors proposed to use a neural network-based strategy for the estimation of an initial condition of the system.

In both cases ([72] and Section 6.2), the authors demonstrated significant advantages that can be gained by using the proposed strategies. However, both of these approaches inherit a common drawback, namely, they are designed for a particular setting of the initial condition of the system. This means that there is no guarantee that the proposed strategies will have the same performance when the initial condition is significantly different than the one used for the design purposes.

The main objective of this section is to present an alternative solution which overcomes the above-mentioned drawbacks. Following the approach described in Section 6.2, let us assume that

$$\boldsymbol{Q}_k = q(\boldsymbol{\varepsilon}_k)\boldsymbol{I} + \delta_1\boldsymbol{I}, \tag{6.32}$$

but $\boldsymbol{R}_k$ is determined by (4.131) and $q(\boldsymbol{\varepsilon}_k)$ is a positive definite function. Contrary to [187], to approximate $q(\boldsymbol{\varepsilon}_k)$, it is suggested to use spline functions [39] instead of genetic programming. Such a choice is not accidental and it is dictated by the computational complexity related to the proposed design procedure, which will be carefully described in the subsequent part of this section. Indeed, many researchers working with genetic programming clearly indicate that its main drawback is related to its computational burden. Similarly, the application of neural networks [96] reduces to the non-linear parameter estimation task, which usually involves a high computational burden. Contrary to neural networks and genetic programming, the suggested spline approximation technique is computationally less demanding.

The solution presented in the subsequent part of this section is based on the general idea of *stochastic robustness method* originally used by Marrison and Stengel [112] for designing robust control systems. In other words, stochastic robustness analysis is a practical method of quantifying the robustness of control systems.

In this work, a stochastic robustness metric characterises an EUIO (denoted by $\mathcal{O}$) with the probability that the observer will have an unacceptable performance in the presence of possible variations of the initial condition of the system $\boldsymbol{x}_0 \in \mathbb{X} \subset \mathbb{R}^n$. The probability P can be defined as the integral of an indicator function over the space of expected variations of the initial condition:

$$P(\boldsymbol{p}) = \int_{\mathbb{X}} I[\mathcal{S}(\boldsymbol{x}_0), \mathcal{O}(\boldsymbol{p})]\mathrm{pr}(\boldsymbol{x}_0)\mathrm{d}\boldsymbol{x}_0, \tag{6.33}$$

where $\mathcal{S}$ stands for the system structure, $\boldsymbol{p}$ denotes the design parameter vector of the observer $\mathcal{O}$, i.e., the parameter vector of $q(\boldsymbol{\varepsilon}_k)$ in (6.32), and $\mathrm{pr}(\boldsymbol{x}_0)$ is the probability density function. Moreover, the binary indicator function is defined as follows:

$$I[\mathcal{S}(\boldsymbol{x}_0), \mathcal{O}(\boldsymbol{p})] = \begin{cases} 0 & \|\boldsymbol{e}_T\|_2 \le \eta\|\boldsymbol{e}_0\|_2, \\ 1 & \text{otherwise}, \end{cases} \tag{6.34}$$

with $\boldsymbol{e}_k = \boldsymbol{x}_k - \hat{\boldsymbol{x}}_k$, while $T > 0$ and $0 < \eta < 1$ are the parameters set by the designer. Thus, by considering (6.34), it can be seen that performance of the observer $\mathcal{O}$ for $\boldsymbol{x}_0$ is acceptable when $\|\boldsymbol{e}_k\|_2$, $k = 0, \ldots, T$, decreases in the time $T$ such that $\frac{\|\boldsymbol{e}_T\|_2}{\|\boldsymbol{e}_0\|_2} \leq \eta$.

Unfortunately, (6.33) cannot be integrated analytically. A practical alternative is to use the Monte Carlo Evaluation (MCE) with $\mathrm{pr}(\boldsymbol{x}_0)$ shaping the random values of $\boldsymbol{x}_0$ denoted by $\boldsymbol{x}_0^i$. When $N$ random $\boldsymbol{x}_0^i$, $i = 1, \ldots, N$ are generated, then the estimate of $P$ can be given as

$$\hat{P}(\boldsymbol{p}) = \frac{1}{N} \sum_{i=1}^{N} I[\mathcal{S}(\boldsymbol{x}_0^i), \mathcal{O}(\boldsymbol{p})], \tag{6.35}$$

while $\hat{P}$ approaches $P$ in the limit as $N \to \infty$. Moreover, it should be pointed out that the initial condition of the observer remains fixed for all $\boldsymbol{x}_0^i$, $i = 1, \ldots, N$, and it should be set by the designer.

It is obvious that it is impossible to set $N = \infty$. Thus, the problem is to select $N$ in such a way as to obtain $\hat{P}$ (which is a random variable) with the standard deviation $\sigma_{\hat{P}}$ less than a predefined threshold. Since the stochastic metric (6.34) is binary, then it is clear that $\hat{P}$ has the binomial distribution with

$$\sigma_{\hat{P}} = \sqrt{\frac{P(\boldsymbol{p}) - P(\boldsymbol{p})^2}{N}}. \tag{6.36}$$

Since $P$ is unknown, then the only feasible way is to calculate the upper bound of (6.36), which gives

$$\sigma_{\hat{P}} \leq \frac{1}{2} \frac{1}{\sqrt{N}}. \tag{6.37}$$

Using (6.37), it can be shown that $N$ can be calculated as follows:

$$N = \left\lceil \frac{1}{4} \sigma_{\hat{P}}^{-2} \right\rceil, \tag{6.38}$$

where $\lceil \cdot \rceil$ is a rounding operator returning an integer value that is not smaller than its argument. Since all the ingredients of the algorithm are given, then its outline can be presented:

Step 1: Select $T$, $\eta$, $\hat{\boldsymbol{x}}_0$ and $\sigma_{\hat{P}}$.
Step 2: Calculate $N$ according to (6.38).
Step 3: Determine $q(\boldsymbol{\varepsilon}_k)$ (by estimating $\boldsymbol{p}$) in (6.32) in such a way as to minimise the cost function (6.35).

In the literature, there are two commonly used ways to represent a polynomial spline: the piecewise polynomial function (pp-form) and its irredundant representation, the so-called basis-splines (B-form) [39]. Given the knot sequence $t_x = \{t_1, t_2, \ldots, t_n\}$, the *B-form* that describes an univariate spline $f(x)$ can be expressed as a weighted sum:

$$f(x) = \sum_{i=1}^{n} c_i B_{i,k}(x), \tag{6.39}$$

where $B_{i,k}(x)$ is a non-negative piecewise-polynomial function of the degree $k$, which is non-zero only on the interval $[t_i, t_{i+k}]$. The simplest method of obtaining multivariate interpolation and approximation routines is to take univariate methods and form a multivariate method via tensor products [39]:

$$f(\boldsymbol{x}) = \sum_{i_1=1}^{N_1} \sum_{i_2=1}^{N_2} \dots \sum_{i_n=1}^{N_n} c_{i_1,i_2,\dots,i_n} B_{i_1,k_1,t_1}(x_1) B_{i_2,k_2,t_2}(x_2) \dots$$
$$\dots B_{i_n,k_n,t_n}(x_n), \tag{6.40}$$

and each variable $x_i$ possesses its own knot sequence $t_i$ of the length $N_i$ and order $k_i$. The problem of determining the coefficients $c_{i_1,i_2,\dots,i_n}$ can be solved efficiently by repeatedly solving univariate interpolation problems, as described in [39]. Thus, (6.40) is used to represent $q(\varepsilon_k)$ in (6.32), i.e., $q(\varepsilon_k) = f(\varepsilon_k)$, and the parameter vector $\boldsymbol{p}$ of $q(\cdot)$ consists of all coefficients $c_{i_1,i_2,\dots,i_n}$, $i_1 = 1, \dots, N_1$, $\dots, i_n = 1, \dots, N_n$.

Unfortunately, (6.35) cannot be differentiated with respect to $\boldsymbol{p}$ and hence the gradient-based algorithm cannot be employed. Indeed, the calculation of (6.35) involves $N$ runs of the algorithm (4.97)–(4.101). Moreover, the cost function (6.35) can be even multi-modal and hence global optimisation techniques should be preferred rather than local optimisation tools. In this work, it is proposed to use the Evolutionary Search with Soft Selection (ESSS) algorithm [63]. The ESSS algorithm is based on probably the simplest selection-mutation model of the Darwinian's evolution [63, 88, 127]. It should be stressed here that the original algorithm is improved by applying a directional mutation [145], and hence the abbreviation $ESSS_\alpha - DM$ is employed in the subsequent part of the work.

As can be noticed in many excellent monographs on evolutionary algorithms (see, e.g., [103, 115]), evolution is a motion of individuals in the phenotype space, called also the adaptation landscape. This motion is caused by the selection and mutation processes. Selection leads to the concentration of individuals around the best ones, but mutation introduces the diversity of phenes and disperses the population in the landscape. The directional distribution [41, 111], introduced in the mutation operator, is used to produce new candidate solutions according to the formula

$$\boldsymbol{p}_{k+1} = \boldsymbol{p}_k + r\,\boldsymbol{d}_k, \tag{6.41}$$

where $\boldsymbol{p}_k$ is the estimate of the parameter vector $\boldsymbol{p}$ in the $k$th iteration of the $ESSS_\alpha - DM$ algorithm, $r$ denotes symmetric $\alpha$-stable variate $r \sim \chi_{\alpha,\gamma} \stackrel{d}{=} |S_\alpha S(\gamma)|$, and $\boldsymbol{d}_k \sim \mathcal{M}(\boldsymbol{\mu}_k, \kappa)$ denotes pseudo-random vector from the directional distribution. The choice of the symmetric $\alpha$-stable distribution is not an accidental one. In the recent years, the class of $S_\alpha S(\gamma)$ distributions has received an increasing interest of the evolutionary computation community [74, 127, 107]. The family of the symmetric stable distribution $S_\alpha S(\gamma)$ is characterised by two parameters: the stable index $\alpha$, which defines the shape of its p.d.f., and the scale $\gamma$. Bearing in mind the fact that evolutionary algorithms are not convergent to the exact optimal solution but to a close area around it [88], the class of the $S_\alpha S(\gamma)$ distribution allows establishing a well-balanced compromise between two mutually exclusive properties: the accuracy of locating potential solution and the ability of escaping from the local optima – two most challenging problems of stochastic optimisation [74, 127].

The directional distribution [41, 111] is determined on the surface of an $n$-dimensional ball; therefore, it is responsible for choosing the direction of mutation. Moreover, $\mathcal{M}(\boldsymbol{\mu}_k, \kappa)$ distributions are parameterised by the pair: the mean direction $\boldsymbol{\mu}$, which defines the most frequently chosen direction, and the concentration parameter $\kappa$, which controls the degree of dispersion around the vector $\boldsymbol{\mu}$. It should be stressed that the value of the concentration parameter strongly depends on the dimension of the search space and becomes constant during the evolution. In order to estimate the vector $\boldsymbol{\mu}$, which aims at increasing the convergence rate of the algorithm, the approach proposed by Obuchowicz [125] is utilised:

$$\boldsymbol{\mu}^t = \frac{\langle \boldsymbol{p}^t \rangle - \langle \boldsymbol{p}^{t-1} \rangle}{\| \langle \boldsymbol{p}^t \rangle - \langle \boldsymbol{p}^{t-1} \rangle \|}, \quad \text{where} \quad \langle \boldsymbol{p}^t \rangle = \frac{1}{\eta} \sum_{k=1}^{\eta} \boldsymbol{p}_k^t. \tag{6.42}$$

To meet the requirement for positive definiteness of the function $q(\boldsymbol{\varepsilon}_k)$, the evolutionary algorithm is supplied with an auxiliary correction algorithm. Let us notice that the function (6.40) used to approximate $q(\boldsymbol{\varepsilon}_k)$ is positive if all spline coefficients $c_{i_1,i_2,\ldots,i_n}$ are non-negative. Therefore, the purpose of the correction procedure is to maintain each individual $\boldsymbol{p}$ in a feasible set. In the theory of evolutionary computation, many techniques for dealing with constraints can be found [115], while in this work a simple reflection procedure is applied:

$$p_{i,\text{new}} = |p_{i,k+1}|, \quad i = 1, \ldots, n_p, \tag{6.43}$$

where $\boldsymbol{p}_{k+1}$ is a new candidate solution obtained as a result of the mutation procedure (6.41). The complete $ESSS_\alpha - DM$ algorithm is given in Tab. 6.2.

## 6.4 Experimental Result

The objective of this section is to determine the effectiveness of the approaches described in Sections 6.1, 6.2 and 6.3 based on experimental results. In particular, two non-linear systems are considered, namely, the induction motor presented in Section 4.1.3 and the valve actuator [12].

### 6.4.1 State Estimation and Fault Diagnosis

Let us reconsider the example with the induction motor presented in Section 4.1.3. Using the same parameters and settings, the following three cases concerning the selection of $\boldsymbol{Q}_{k-1}$ and $\boldsymbol{R}_k$ in the EUIO algorithm described in Section 4.1 were considered:

Case 1: Classic approach (constant values), i.e., $\boldsymbol{Q}_{k-1} = 0.1\boldsymbol{I}$, $\boldsymbol{R}_k = 0.1\boldsymbol{I}$,

**Table 6.2.** Description of the $ESSS_\alpha - DM$ algorithm

***Input data***

$n_m$ – population size;

$t_{\max}$ – maximum number of iterations (epochs);

$\gamma, \alpha, \kappa$ – parameters of mutation: scale, stable index and concentration;

$\hat{P} : \mathbb{R}^n \to \mathbb{R}_+$ – fitness function defined by (6.35);

$\hat{\boldsymbol{p}}_0^0$ – initial estimate.

***Algorithm***

1. Initialise

$$G(0) = \left(\hat{\boldsymbol{p}}_1^0, \hat{\boldsymbol{p}}_2^0, \dots, \hat{\boldsymbol{p}}_{n_m}^0\right), \quad \hat{\boldsymbol{p}}_k^0 = \hat{\boldsymbol{p}}_0^0 + \boldsymbol{s},$$

where $\boldsymbol{s} \sim \mathcal{N}(0, \gamma \boldsymbol{I}_n), \quad k = 1, 2, \dots, n_m$

2. Repeat

(a) *Estimation*

$$\Phi\big(G(t)\big) = \left(\phi_1^t, \phi_2^t, \dots, \phi_{n_m}^t\right), \quad \text{where} \quad \phi_k^t = \hat{P}\big(\hat{\boldsymbol{p}}_k^t\big), \quad k = 1, 2, \dots, n_m.$$

(b) *Proportional selection*

$$G(t) \longrightarrow G(t)' = \left(\hat{\boldsymbol{p}}_{h_1}^t, \hat{\boldsymbol{p}}_{h_2}^t, \dots, \hat{\boldsymbol{p}}_{h_{n_m}}^t\right),$$

$$\big(h_1, h_2, \dots, h_{n_m}\big), \text{ where } h_k = \min\left\{h : \frac{\sum_{l=1}^{h} \phi_l^t}{\sum_{l=1}^{n_m} \phi_l^t} > \zeta_k\right\}$$

and $\{\zeta_k\}_{k=1}^{n_m}$ are random numbers uniformly distributed in $[0, 1)$.

(c) *Estimation of the most promising direction of mutation*

$$\boldsymbol{\mu}(t) = \frac{\langle \hat{\boldsymbol{p}}^t \rangle - \langle \hat{\boldsymbol{p}}^{t-1} \rangle}{\| \langle \hat{\boldsymbol{p}}^t \rangle - \langle \hat{\boldsymbol{p}}^{t-1} \rangle \|}, \quad \text{where} \quad \langle \hat{\boldsymbol{p}}^t \rangle = \frac{1}{n_m} \sum_{k=1}^{n_m} \hat{\boldsymbol{p}}_{h_k}^t$$

(d) *Mutation*

$$G(t)' \longrightarrow G(t)'';$$

$$\hat{\boldsymbol{p}}_k^{t+1} = \hat{\boldsymbol{p}}_{h_k}^t + \chi_{\alpha,\gamma} \boldsymbol{u}^{(n)}, \; \boldsymbol{u}^{(n)} \sim \mathcal{M}(\boldsymbol{\mu}(t), \kappa), \; k = 1, 2, \dots, n_m.$$

(e) *Correction algorithm*

$$G(t)'' \longrightarrow G(t+1);$$

Until $t > t_{\max}$.

---

Case 2: Selection according to (4.59), i.e.

$$\boldsymbol{Q}_{k-1} = 10^3 \boldsymbol{\varepsilon}_{k-1}^T \boldsymbol{\varepsilon}_{k-1} \boldsymbol{I} + 0.01 \boldsymbol{I}, \; \boldsymbol{R}_k = 10 \boldsymbol{\varepsilon}_k^T \boldsymbol{\varepsilon}_k \boldsymbol{I} + 0.01 \boldsymbol{I}, \tag{6.44}$$

Case 3: GP-based approach presented in Section 6.2.

It should be pointed out that in all the cases the unknown input-free mode (i.e., $\boldsymbol{d}_k = \boldsymbol{0}$) is considered. This is because the main purpose of this example is to show the importance of an appropriate selection of instrumental matrices but not the abilities of disturbance decoupling.

In order to obtain the matrices $\boldsymbol{Q}_{k-1}$ and $\boldsymbol{R}_k$ using the GP-based approach (Case 3), a set of $n_t = 300$ input-output measurements was generated according to (4.60)–(4.65), and then the approach from Section 6.2 was applied. As a result, the following form of instrumental matrices was obtained:

$$\boldsymbol{Q}_k = \left(10^2\varepsilon_{1,k}^2\varepsilon_{2,k}^2 + 1012\varepsilon_{1,k} + 103.45\varepsilon_{1,k} + 0.01\right)^2 \boldsymbol{I}, \tag{6.45}$$

$$\boldsymbol{R}_k = \left(112\varepsilon_{1,k}^2 + 0.1\varepsilon_{1,k}\varepsilon_{2,k} + 0.12\right)^2 \boldsymbol{I}. \tag{6.46}$$

The parameters used in the GP algorithm presented in Section 6.2 were $n_m = 200$, $n_d = 10$, $n_s = 10$, $\mathbb{F} = \{+, *, /\}$. It should also be pointed out that the above matrices (6.45)–(6.46) are formed by simple polynomials. This, however, may not be the case for other applications.

The simulation results (for all the cases) are shown in Fig. 6.6. The numerical values of the optimisation index (6.31) are as follows: Case 1 $j_{\text{obs},3} = 1.49 \cdot 10^5$, Case 2 $j_{\text{obs},3} = 1.55$, Case 3 $j_{\text{obs},3} = 1.2 \cdot 10^{-16}$. Both the above results and the plots shown in Fig. 6.6 confirm the relevance of an appropriate selection of the instrumental matrices. Indeed, as can be seen, the proposed approach is superior to the classic technique of selecting the instrumental matrices $\boldsymbol{Q}_{k-1}$ and $\boldsymbol{R}_k$.

Let us reconsider the example with an induction motor presented in Section 4.1.3. This time, the EUIO described in Section 4.2 is employed and its convergence is enhanced with the approach described in Section 6.3.

Let $\mathbb{X}$ be a bounded hypercube denoting the space of the possible variations of the initial condition $\boldsymbol{x}_0$:

$$\mathbb{X} = \{[-276, 279] \times [-243, 369] \times [-15, 38] \times [-11, 52] \times [-11, 56]\} \subset \mathbb{R}^5. \tag{6.47}$$

Let us assume that each initial condition of the system $\boldsymbol{x}_0$ is equally probable, i.e.,

$$pr(\boldsymbol{x}_0) = \begin{cases} \frac{1}{m(\mathbb{X})} & \text{for } \boldsymbol{x}_0 \in \mathbb{X} \\ 0 & \text{otherwise} \end{cases},$$

where $m(\mathbb{A})$ is the Lebesgue measure of the set $\mathbb{A}$.

First, let us start with the unknown input-free case, i.e., $\boldsymbol{E}_k = \boldsymbol{0}$. In order to completely define the indicator function (6.34), the following values of its parameters were chosen: $T = 30$ and $\eta = 0.001$. This means that the main objective is to obtain the design parameter vector $\boldsymbol{p}$ in such a way as to minimise the probability (estimated by (6.35)) that the observer $\mathcal{O}$ will not converge in such a way that $\frac{\|\boldsymbol{e}_{30}\|_2}{\|\boldsymbol{e}_0\|_2} \leq 0.001$ for any initial condition $\boldsymbol{x}_0 \in \mathbb{X}$, where the initial condition for the observer is given by $\hat{\boldsymbol{x}}_0 = [1.5, 63, 11.5, 20.5, 22.5]^T$, which is the centre of $\mathbb{X}$.

The standard deviation of $\hat{P}(\boldsymbol{p})$ was selected as $\sigma_{\hat{P}} = 0.005$ and hence, according to (6.38), $N = 10000$. As can be observed from (4.65), the dimension of $\boldsymbol{\varepsilon}_k$

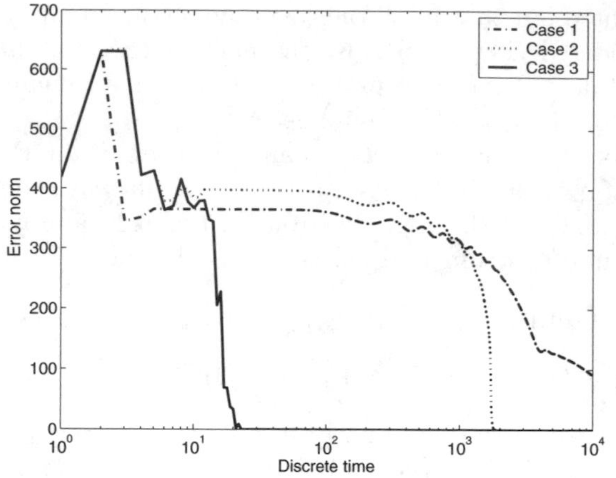


**Fig. 6.6.** State estimation error norm $\|\boldsymbol{e}_k\|_2$ for Case 1, Case 2 and Case 3

is $m = 2$. Thus, the two-dimensional spline function of the degree $k_{x_1} = k_{x_2} = 3$ used in the experiment to approximate $q(\boldsymbol{\varepsilon}_k)$ is defined by the knot sequence $t_{x_1} = t_{x_2} = \{-200 + (i-1)\frac{400}{19}\}_{i=1}^{20}$; consequently, $N_1 = N_2 = 20$.

The parameters related to the $ESSS_\alpha - DM$ algorithm were selected as follows: the initial parameter vector $\hat{\boldsymbol{p}}_0^0$ was randomly generated, the population size was $n_m = 50$, the maximum number of iterations $t_{\max} = 1000$, $\gamma = 0.1$, $\alpha = 1.5$, and $\kappa = 0.01$. Finally, the instrumental matrix $\boldsymbol{R}_k$ was $\boldsymbol{R}_k = 10^{-3}\boldsymbol{I}$.

As a result of using the $ESSS_\alpha - DM$ algorithm for the minimisation of (6.35) with respect to the design parameter vector $\boldsymbol{p}$, it was determined that $\hat{P} = 0.0819$.

The shape of the resulting function $q(\boldsymbol{\varepsilon}_k)$ is presented in Fig. 6.7. In particular, Fig. 6.7b presents its shape in the whole domain of $\boldsymbol{\varepsilon}_k$ being considered, while Fig. 6.7a exhibits its small part around $\boldsymbol{\varepsilon}_k = \boldsymbol{0}$. It can be observed in Fig. 6.7a that the value of $q(\boldsymbol{\varepsilon}_k)$ increases rapidly when $\boldsymbol{\varepsilon}_k$ starts to diverge from $\boldsymbol{0}$, which is consistent with the theoretical analysis performed in Section 4.2. For the sake of comparison, the selection strategy of $\boldsymbol{Q}_k$ proposed in [19] was employed, i.e., $\boldsymbol{Q}_k = 10^3\boldsymbol{\varepsilon}_k^T\boldsymbol{\varepsilon}_k\boldsymbol{I} + 10^{-3}\boldsymbol{I}$. As a result, it was figured out that $\hat{P} = 0.8858$. This means that for the total number of $N = 10000$ initial conditions, the observer considered cannot provide an acceptable performance for 8858 cases. In the case of the proposed observer there are 819 unacceptable cases, which is definitely a better result. This situation is clearly exhibited in Fig. 6.8 (successful runs are denoted by the dark colour), which shows the trajectories of $\|\boldsymbol{e}_k\|_2$ (for all $N = 10000$ cases) for the proposed observer (Fig. 6.8a) and the observer described in [19] (Fig. 6.8b).

### Unknown input decoupling and fault detection

The objective of presenting the next example is to show the abilities of unknown input decoupling. First, let us assume that the unknown input distribution matrix is

$$E = [1.2,\ 0.2,\ 2.4,\ 1,\ -1.6]^T. \tag{6.48}$$

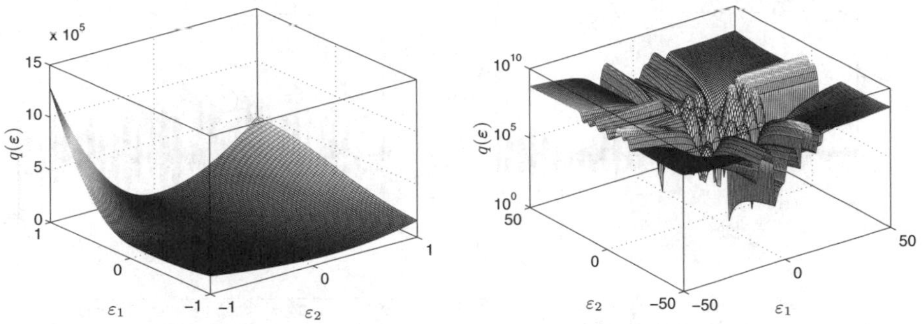


(a) (b)

**Fig. 6.7.** Shape of the obtained $q(\boldsymbol{\varepsilon}_k)$ (a) and its small part around $\boldsymbol{\varepsilon}_k = \mathbf{0}$ (b)

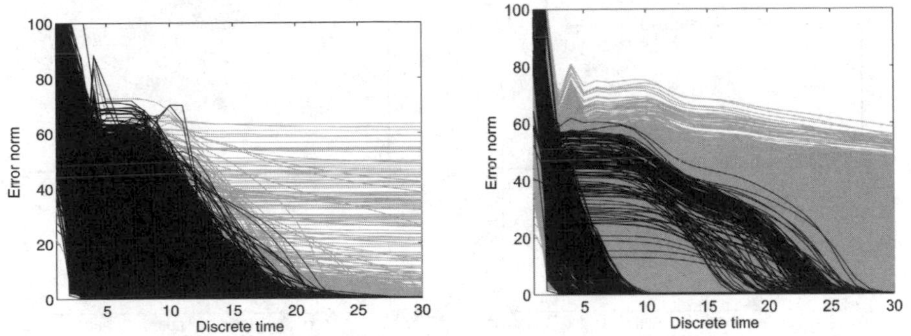


(a) (b)

**Fig. 6.8.** Trajectories of $\|\boldsymbol{e}_k\|_2$ for the proposed observer (a) and the observer described in [19] (b)

Thus, the system (4.69)–(4.70) is described using (4.60)–(4.65) and (6.48). Since the system description is given, then it is possible to design the extended unknown input observer in the same way and with the same parameter values as in Section 4.1. Because $\boldsymbol{d}_k$ is unknown it is impossible to use it in the design procedure. On the other hand, only the knowledge regarding (6.48) is necessary and hence any form of $\boldsymbol{d}_k$ can be used for design purposes. In this section, the following setting is used: $\boldsymbol{d}_k = \mathbf{0}$.

As a result of using the proposed approach, the function $q(\boldsymbol{\varepsilon}_k)$ was obtained for which $\hat{P} = 0.0834$. Since the observer is designed, it is possible to check its performance with respect to unknown input decoupling. For that purpose, let us assume that the unknown input is given by

$$d_k = 3.0\sin(0.5\pi k)\cos(0.03\pi k). \tag{6.49}$$

Figure 6.9 presents the residual $\boldsymbol{z}_k$ for the observer designed in Section 4.2. From this figure, it is clear that the unknown input influences the residual signal

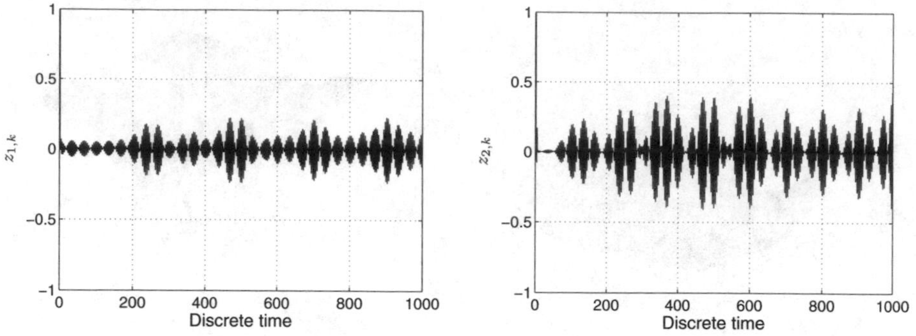


(a) (b)

**Fig. 6.9.** Residuals for an observer without unknown input decoupling

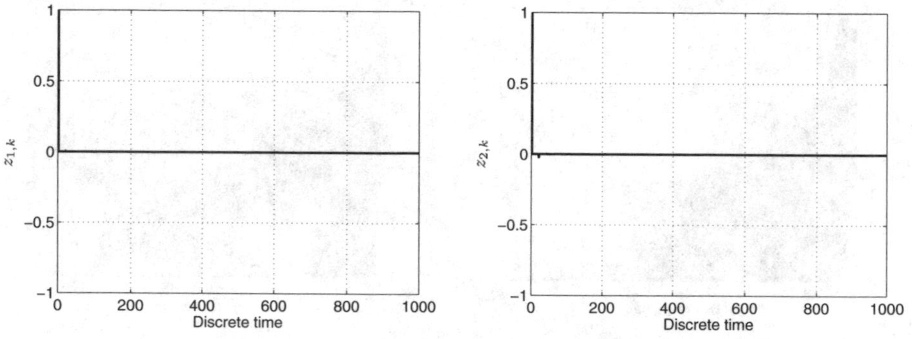


(a) (b)

**Fig. 6.10.** Residuals for an observer with unknown input decoupling

and hence it may cause an unreliable fault detection (and, consequently, fault isolation). On the contrary, Fig. 6.10 shows the residual with unknown input decoupling. In this case the residual is almost zero, which confirms the importance of unknown input decoupling. The objective of presenting the next example is to show the effectiveness of the proposed observer as a residual generator in the presence of an unknown input. For that purpose, the following fault scenarios were considered:

Case 1: Abrupt fault of the $y_{1,k}$ sensor:

$$f_{1,k} = \begin{cases} 0, & 500 < k < 140, \\ -0.1 y_{1,k}, & \text{otherwise}, \end{cases} , \tag{6.50}$$

and $f_{2,k} = 0$.

Case 2: Abrupt fault of the $u_{1,k}$ actuator:

$$f_{2,k} = \begin{cases} 0, & 500 < k < 140, \\ -0.2u_{1,k}, & \text{otherwise.} \end{cases}, \tag{6.51}$$

and $f_{1,k} = 0$.

Thus, the system is now described by (4.24)–(4.25) with (4.60)–(4.65), (6.48), $\boldsymbol{f}_k = [f_{1,k}, f_{2,k}]^T$, and

$$\boldsymbol{L}_{1,k} = \begin{bmatrix} \frac{1}{\sigma L_s} & 0\,0\,0\,0 \\ 0 & 0\,0\,0\,0 \end{bmatrix}^T, \tag{6.52}$$

$$\boldsymbol{L}_{2,k} = \begin{bmatrix} 1 & 0 \\ 0 & 0 \end{bmatrix}. \tag{6.53}$$

From Figs. 6.11 and 6.12, it can be observed that the residual signal is sensitive to the faults under consideration. This, together with unknown input decoupling, implies that the process of fault detection becomes a relatively easy task.

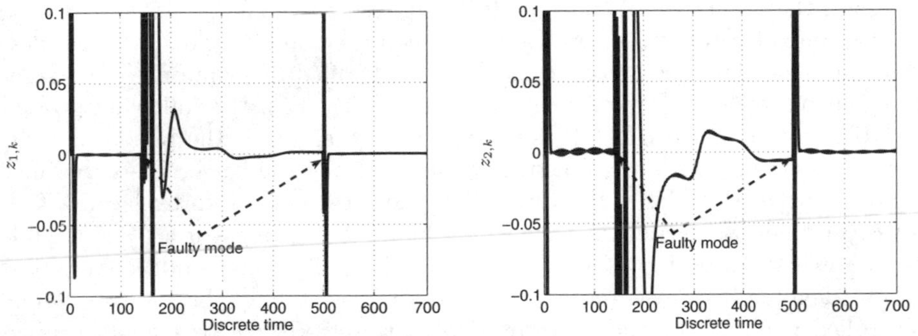


(a) (b)

**Fig. 6.11.** Residuals for a sensor fault

### 6.4.2 Industrial Application

This section presents an industrial application study regarding the techniques presented in Sections 6.1 and 6.2. In particular, the presented example concerns fault detection of a valve actuator with state-space models designed with GP and the EUIO. The problem regarding FDI of this actuator was attacked from different angles in the EU DAMADICS project (see, e.g., the examples of Chapter 7). DAMADICS (*Development and Application of Methods for Actuator Diagnosis in Industrial Control Systems*) was a research project focused on drawing together wide-ranging techniques and fault diagnosis within the

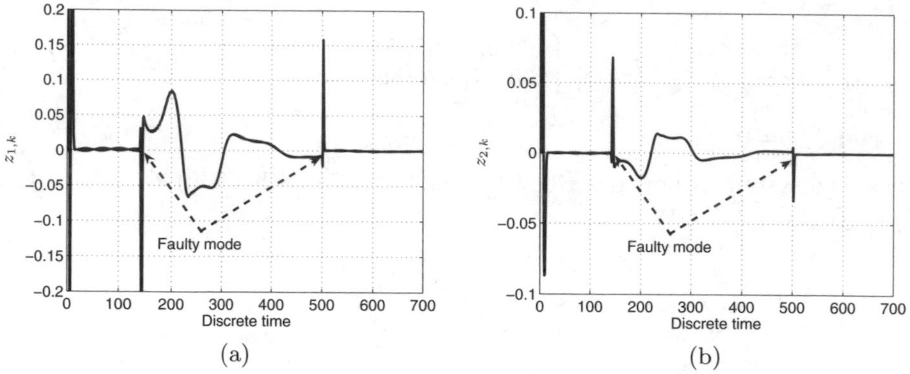


**Fig. 6.12.** Residuals for an actuator fault

framework of a real application to on-line diagnosis of a 5-stage evaporisation plant of the sugar factory in Lublin, Poland. The project was focused on the diagnosis of valve (cf. Fig. 6.13) plant actuators and looked towards real implementation methods for new actuator systems. The sugar factory was a subcontractor (under the Warsaw University of Technology) providing real process data and the evaluation of trials of fault diagnosis methods.

The control valve is a mean used to prevent, permit and/or limit the flow of sugar juice through the control system (a detailed description of this actuator can be found in [36]). As can be seen in Fig. 6.13, the following process variables can be measured: $CV$ is the control signal, $P1$ is the pressure at the inlet of the valve, $P2$ is the pressure at the outlet of the valve, $T1$ is the juice temperature at the inlet of the valve, $X$ is the servomotor rod displacement, $F$ is the juice flow at the outlet of the valve. Thus, the output is $\boldsymbol{y} = (F, X)$, while the input is given by $\boldsymbol{u} = (CV, P1, P2, T)$. In Fig. 6.13, three additional bypass valves (denoted by $z_1$, $z_2$, and $z_3$) can be seen. The state of these valves can be controlled manually by the operator. They are introduced for manual process operation, actuator maintenance and safety purposes. The data gathered from the real plant can be found on the DAMADICS website [36]. Although a large amount of real data is available, they do not cover all faulty situations, while the simulator is able to generate a set of 19 faults (see Tab. 6.3) Moreover, due to a strict production regime, operators do not allow changing plant inputs, i.e., they are set up by control systems. Thus, an actuator simulator was developed with MATLAB Simulink (available on [36]). Apart from experimental design purposes, the main reason for using the data from the simulator is the fact that the achieved results can be easily compared with the results achieved with different approaches, e.g., [114, 128, 183]. The objective of this section is to design the state-space model of the actuator being considered according to the approach described in Section 6.1.2. The parameters used during the identification process were $n_m = 200$, $n_d = 10$, $n_s = 10$, $\mathbb{F} = \{+, *, /\}$. For the sake of comparison,

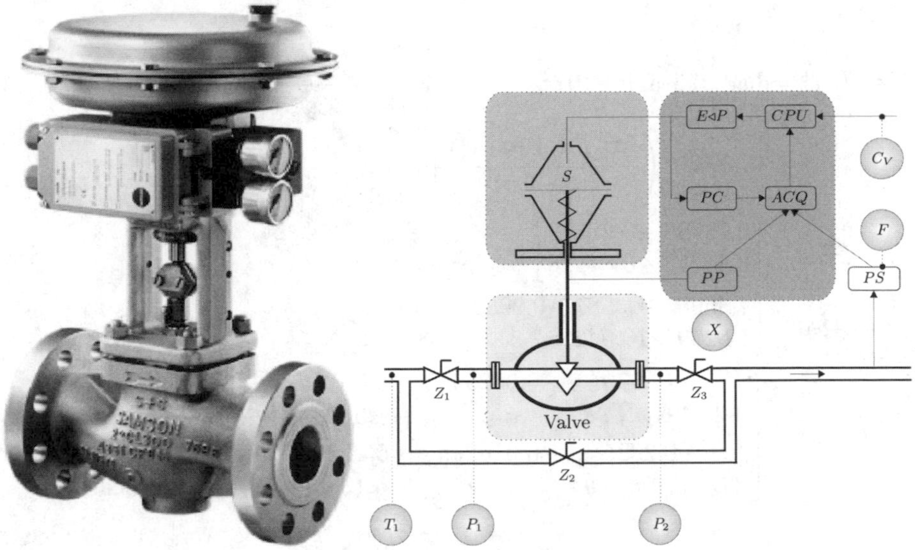


**Fig. 6.13.** Actuator and its scheme

the linear state-space model was obtained with the use of the MATLAB System Identification Toolbox. In both the linear and non-linear cases, the order of the model was tested between $n = 2, \ldots, 8$. Unfortunately, the relation between the input $\boldsymbol{u}_k$ and the juice flow $y_{1,k}$ cannot be modelled by a linear state-space model. Indeed, the modelling error was approximately 35%, thus making the linear model unacceptable. On the other hand, the relation between the input $\boldsymbol{u}_k$ and the rod displacement $y_{2,k}$ can be modelled, with very good results, by the linear state-space model. Bearing this in mind, the identification process was decomposed into two phases, i.e.,

1. Derivation of a relation between the rod displacement and the input with a linear state-space model.
2. Derivation of a relation between the juice flow and the input with a non-linear state-space model designed by GP.

Experimental results showed that the best-suited linear model is of order $n = 2$. After 50 runs of the GP algorithm performed for each model order, it was found that the order of the model which provides the best approximation quality is $n = 2$. Thus, as a result of combining both the linear and non-linear models, the following model structure of an actuator was obtained:

$$\boldsymbol{x}_{k+1} = \begin{bmatrix} \boldsymbol{A}_F(\boldsymbol{x}_k) & 0 \\ 0 & \boldsymbol{A}_X \end{bmatrix} \boldsymbol{x}_k + \begin{bmatrix} \boldsymbol{h}(\boldsymbol{u}_k) \\ \boldsymbol{B}_X \boldsymbol{u}_k \end{bmatrix}, \tag{6.54}$$

$$\boldsymbol{y}_{M,k+1} = \boldsymbol{C}\boldsymbol{x}_{k+1}, \tag{6.55}$$

where

$$\begin{aligned}\boldsymbol{A}_F(\boldsymbol{x}_k) =&\text{diag}\left[0.3\tanh\left(10x_{1,k}^2 + 23x_{1,k}x_{2,k} + \frac{26x_{1,k}}{x_{2,k}+0.01}\right),\right.\\ &\left.0.15\tanh\left(\frac{5x_{2,k}^2 + 1.5x_{1,k}}{x_{1,k}^2+0.01}\right)\right],\\ \boldsymbol{A}_X =& \begin{bmatrix} 0.78786 & -0.28319 \\ 0.41252 & -0.84448 \end{bmatrix},\\ \boldsymbol{B}_X =& \begin{bmatrix} 2.3695 & -1.3587 & -0.29929 & 1.1361 \\ 12.269 & -10.042 & 2.516 & 0.83162 \end{bmatrix},\\ \boldsymbol{h}(\boldsymbol{u}_k) =& \begin{bmatrix} -1.087u_{1,k}^2 + 0.0629u_{2,k}^2 - 0.5019u_{3,k}^2 - 3.0108u_{4,k}^2 \\ +0.9491(u_{1,k}u_{2,k} - u_{1,k}u_{3,k}) - 0.5409\frac{u_{1,k}u_{4,k}}{u_{2,k}u_{3,k}+0.01} + 0.9783 \\ -0.292u_{1,k}^2 + 0.0162u_{2,k}^2 - 0.1289u_{3,k}^2 - 0.7733u_{4,k}^2 \\ +0.2438(u_{1,k}u_{2,k} - u_{1,k}u_{3,k}) - 0.1389\frac{u_{1,k}u_{4,k}}{u_{2,k}u_{3,k}+0.01} + 0.2513 \end{bmatrix},\end{aligned}$$

$$\boldsymbol{C} = \begin{bmatrix} 1 & 1 & 0 & 0 \\ 0 & 0 & 0.79 & -0.047 \end{bmatrix}.$$

The mean-squared output error for the model (6.54)–(6.55) was 0.0079. The response of the model obtained for the validation data set is given in Fig. 6.14. From these results, it can be seen that the proposed non-linear state-space model identification approach can effectively be applied to various system identification tasks. The main differences between the behaviour of the model and the system can be observed (cf. Fig. 6.14) for the non-linear model (juice flow) during the saturation of the system. This inaccuracy constitutes the main part of modelling uncertainty.

The main drawback to the GP-based identification algorithm concerns its convergence abilities. Indeed, it seems very difficult to establish conditions which can guarantee the convergence of the proposed algorithm. On the other hand, many examples treated in the literature, cf. [47, 69, 103] and the references therein, as well as the author's experience with GP [114, 183, 187] confirm its particular usefulness, in spite of the lack of the convergence proof. Based on the fitness attained by each of the 50 models resulting from the 50 runs, it is possible to obtain a histogram representing the achieved fitness values (Fig. 6.15) as well as the fitness confidence region. Let $\alpha = 0.99$ denote the confidence level. Then the corresponding confidence region can be defined as

$$\bar{J}_m \in [\bar{j}_m - t_\alpha \frac{s}{\sqrt{50}}, \bar{j}_m + t_\alpha \frac{s}{\sqrt{50}}], \tag{6.56}$$

where $\bar{j}_m = 0.0097$ and $s = 0.0016$ denote the arithmetic mean and standard deviation of the fitness of the 50 models, respectively, while $t_\alpha = 2.58$ is the normal distribution quantile. According to (6.56), the fitness confidence region is $\bar{J}_m \in [0.0093, 0.0105]$, which means that the probability that the true mean fitness $\bar{J}_m$ belongs to this region is 99%. These results confirm that the proposed approach can effectively be used for designing non-linear state-space models,

**Table 6.3.** Set of faults considered for the benchmark (abrupt faults: S – small, M – medium, B – big, I – incipient faults)

| Fault | Description | S | M | B | I |
|---|---|---|---|---|---|
| $f_1$ | Valve clogging | x | x | x | |
| $f_2$ | Valve plug or valve seat sedimentation | | | x | x |
| $f_3$ | Valve plug or valve seat erosion | | | | x |
| $f_4$ | Increased valve or busing friction | | | | x |
| $f_5$ | External leakage | | | | x |
| $f_6$ | Internal leakage (valve tightness) | | | | x |
| $f_7$ | Medium evaporation or critical flow | x | x | x | x |
| $f_8$ | Twisted servomotor's piston rod | x | x | x | |
| $f_9$ | Servomotor housing or terminal tightness | | | | x |
| $f_{10}$ | Servomotor's diaphragm perforation | x | x | x | |
| $f_{11}$ | Servomotor's spring fault | | | | x |
| $f_{12}$ | Electro-pneumatic transducer fault | x | x | x | |
| $f_{13}$ | Rod displacement sensor fault | x | x | x | x |
| $f_{14}$ | Pressure sensor fault | x | x | x | |
| $f_{15}$ | Positioner feedback fault | | | x | |
| $f_{16}$ | Positioner supply pressure drop | x | x | x | |
| $f_{17}$ | Unexpected pressure change across the valve | | | x | x |
| $f_{18}$ | Fully or partly opened bypass valves | x | x | x | x |
| $f_{19}$ | Flow rate sensor fault | x | x | x | |

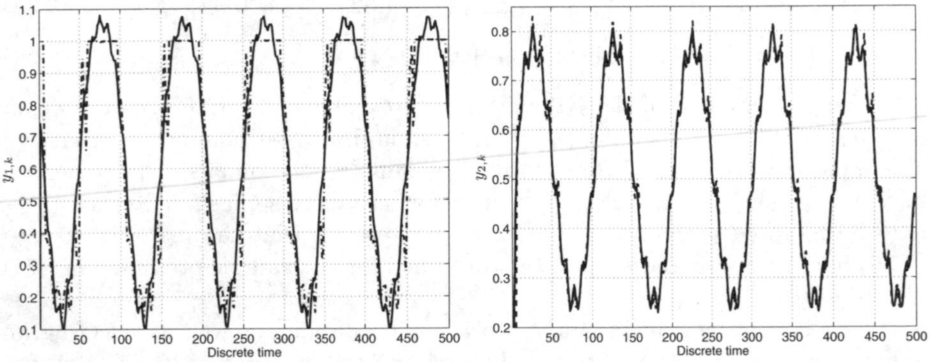


**Fig. 6.14.** System (dotted) and model (solid) outputs (juice flow – left, rod displacement – right) for the validation data set

even without the convergence proof. As the state-space model (6.54)–(6.55) is available, it is possible to determine the unknown input distribution matrix $\boldsymbol{E}_k$ and, consequently, the matrix $\boldsymbol{H}_k$, which is necessary to design the EUIO described in Section 4.1. The details regarding the derivation of the above matrices can be found in [183]. The resulting matrix $\boldsymbol{H}_k$ has the following form:

$$\boldsymbol{H}_k = \begin{bmatrix} 0.2074 \; 0 \; 0 \; 0 \\ 0.3926 \; 0 \; 0 \; 0 \end{bmatrix}. \qquad (6.57)$$

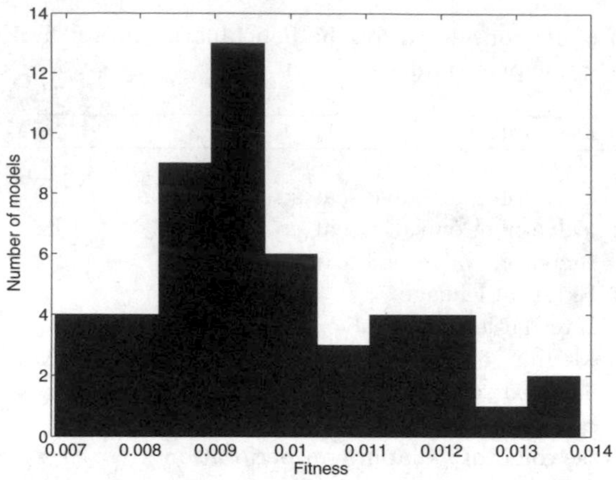


**Fig. 6.15.** Histogram representing the fitness of the 50 models

Since the matrix $\boldsymbol{H}_k$ is known, the instrumental matrices $\boldsymbol{Q}_{k-1}$ and $\boldsymbol{R}_k$ can be determined. In order to obtain the matrices $\boldsymbol{Q}_{k-1}$ and $\boldsymbol{R}_k$ using the GP-based approach, a set of $n_t = 1000$ input-output measurements was utilised, and then the approach from Section 6.2 was applied. As a result, the following form of the instrumental matrices was obtained:

$$\boldsymbol{Q}_{k-1} = \left(10^3\varepsilon_{1,k-1}^2\varepsilon_{2,k-1}^2 + 12\varepsilon_{1,k-1} + 4.32\varepsilon_{1,k-1} + 0.02\right)^2 \boldsymbol{I}, \tag{6.58}$$

$$\boldsymbol{R}_k = \left(112\varepsilon_{1,k}^2 + 0.01\varepsilon_{1,k}\varepsilon_{2,k} + 0.02\right)^2 \boldsymbol{I}. \tag{6.59}$$

The parameters used in the GP algorithm presented in Section 6.2 were $n_m = 300$, $n_d = 10$, $n_s = 10$, $\mathbb{F} = \{+, *, /\}$. It should also be pointed out that the above matrices (6.58)–(6.59) are formed by simple polynomials. This, however, may not be the case for other applications where more sophisticated structures may be required, although it should be pointed out that the matrices (6.58)–(6.59) have a very similar form to that obtained for an example presented in [187] (see (6.45)–6.46)).

As a result of introducing unknown input decoupling as well as selecting an appropriate form of the instrumental matrices $\boldsymbol{Q}_{k-1}$ and $\boldsymbol{R}_k$ for the EUIO, the mean-squared output error was reduced from 0.0079 to 0.0022. As can be seen from Figs. 6.14 and 6.16, all these efforts help to achieve a better modelling quality and lead to more reliable residual generation. Since the EUIO-based residual generator is available, then threshold determination and fault diagnosis can be performed. A detailed description of the above tasks can be found in [183]. In this section, only the final results are presented.

In order to test the designed fault detection scheme, data sets with faults were generated. It should be pointed out that only abrupt faults were considered (cf. Tab. 6.3). Table 6.4 shows the results of fault detection and provides a comparative study between the results achieved by the proposed fault detection scheme

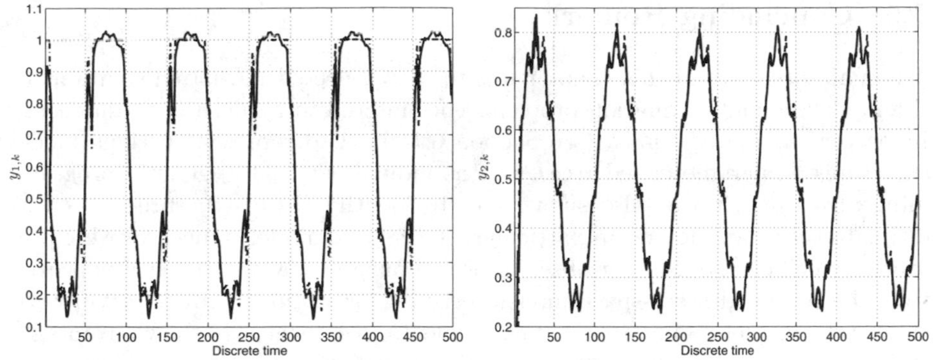


**Fig. 6.16.** System (dotted) and EUIO (solid) outputs (juice flow – left, rod displacement – right)

**Table 6.4.** Results of fault detection (D – detectable, N – not detectable, PD – possible but hard to detect)

| Fault | Description | S | M | B |
|---|---|---|---|---|
| $f_1$ | Valve clogging | D | D(D) | D(D) |
| $f_2$ | Valve plug or valve seat sedimentation | | | D(D) |
| $f_7$ | Medium evaporation or critical flow | D | D(PD) | D(D) |
| $f_8$ | Twisted servomotor's piston rod | N | N(N) | N(N) |
| $f_{10}$ | Servomotor's diaphragm perforation | D | D(PD) | D(D) |
| $f_{11}$ | Servomotor's spring fault | | | D(PD) |
| $f_{12}$ | Electro-pneumatic transducer fault | N | N(PD) | D(PD) |
| $f_{13}$ | Rod displacement sensor fault | D | D(D) | D(D) |
| $f_{15}$ | Positioner feedback fault | | | D(N) |
| $f_{16}$ | Positioner supply pressure drop | N | N(PD) | D(D) |
| $f_{17}$ | Unexpected pressure change across the valve | | | D(PD) |
| $f_{18}$ | Fully or partly opened bypass valves | D | D(D) | D(D) |
| $f_{19}$ | Flow rate sensor fault | D | D(PD) | D(D) |

and the results provided by [163] concerning the qualitative approach to fault detection. The notation given in Tab. 6.4 can be explained as follows: D means that we are 100% that a fault has occurred, N means that it is impossible to detect a given fault, PD means that the fault detection system does not provide enough information to be 100% sure that a fault has occurred.

From Tab. 6.4 it can be seen that it is impossible to detect the fault $f_8$. Indeed, the effect of this fault is exactly at the same level as the effect of noise. The residual is the same as that the for fault-free case, and hence it is impossible to detect this fault. There are also some problems with a few small and/or medium faults; however, it should be pointed out that all faults (except for $f_8$) which are considered as big can be detected.

## 6.5 Concluding Remarks

The main objective of this chapter was to show selected EA-based solutions to the modelling and estimation problems encountered in modern fault diagnosis. In particular, it was shown (see Section 6.1) how to represent various model structures as parameterised trees, i.e., as individuals in the genetic programming algorithm. It was also shown how to identify their structure as well as estimate their parameters. In particular, state-space model structures were presented which can be used for observer design purposes. Moreover, it was proven that the proposed state-space model designing procedures provide asymptotically stable models. The presented experimental results clearly show advantages that can be gained while using the presented approach.

The main drawback of this approach is its computational cost resulting in a relatively long identification time. However, as usual, the model construction procedure is realised off-line, and hence identification time is not very important. Another drawback is that the model order has to be determined by a time-consuming trial-and-error process. This is, however, a problem with all non-linear schemes. There are, of course, some approaches [123] which can be applied to estimate such a model order implicitly.

Sections 6.2 and 6.3 present two different approaches that can be used for increasing the convergence rate of the EUIOs described in Sections 4.1 and 4.2, respectively. It was revealed in Section 4.1 that an appropriate selection of the instrumental matrices $\boldsymbol{Q}_{k-1}$ and $\boldsymbol{R}_k$ strongly influences the convergence properties of the observer. To tackle the instrumental matrices selection problem, a genetic programming-based approach was proposed in Section 6.2. However, it should be pointed out that the proposed technique does not provide a general solution to all problems. Indeed, it makes it possible to obtain a particular form of the instrumental matrices $\boldsymbol{Q}_{k-1}$ and $\boldsymbol{R}_k$ which is only appropriate for the problem being considered. This is, in fact, the main drawback of the proposed approach.

A more general solution was proposed in Section 6.3. To improve the convergence of the EUIO described in Section 4.2, the stochastic robustness technique was utilised to form a stochastic robustness metric describing an unacceptable performance of the EUIO. This section proposes and describes a design procedure that can be used for minimising the probability of such an unacceptable performance. In particular, it was shown that observer performance can be significantly improved with an appropriate selection of the instrumental matrix $\boldsymbol{Q}_k$. For that purpose, the B-spline approximation technique and evolutionary algorithms were utilised.

The main advantage of the proposed EUIO is that its convergence rate is maximised for the whole set of possible initial conditions $\mathbb{X}$ and not only for a single and fixed initial condition as was the case in [19, 187] and in Section 6.2. This superiority was clearly presented with the example regarding state estimation of an induction motor.

Irrespective of the above superiority, the experimental results for the DAMADICS benchmark clearly show that the EUIO described in Section 4.1 is also a valuable FDI tool.

# 7. Neural Networks

In Chapter 4, three different unknown input observer structures were proposed and their design procedures were described in detail. The main assumption underlying the above design procedures was that they should be simple enough to becom useful for engineering applications. Moreover, in Chapter 6 a number of solutions that can be used for increasing the performance of UIO-based FDI schemes were proposed and carefully described.

Apart from the unquestionable effectiveness of the above-mentioned approaches, there are examples for which fault directions are very similar to that of an unknown input. This may lead to a situation in which the effect of some faults is minimised and hence they may be impossible to detect. Other approaches that make use of the idea of an unknown input also inherit these drawbacks, e.g., robust parity relations approaches [27]. An obvious approach to tackle such a challenging problem is to use a different description of model uncertainty. Such a description was discussed in Section 2.3. Instead of decoupling model uncertainty, the knowledge about it is used to form the so-called system output confidence interval (2.47), which can then be used for robust fault diagnosis. It is important to note that the parameters of the model underlying such a fault detection scheme do not necessarily have to have physical meaning. This extends considerably the spectrum of candidate models that can be used for design purposes. The main objective of this chapter is to show how to use artificial neural networks in such a robust fault detection scheme. Contrary to the industrial applications of neural networks that are presented in the literature [27, 71, 89, 152], the task of designing a neural network is defined in this chapter in such a way as to obtain a model with a possibly small uncertainty. Indeed, the approaches presented in the literature try to obtain a model that is best suited to a particular data set. This may result in a model with a relatively large uncertainty. A degraded performance of fault diagnosis constitutes a direct consequence of using such models.

Taking into account the above discussion, the chapter is organised as follows: Section 7.1 extends the general ideas of the experimental design described in Chapter 5 to neural networks. In particular, one objective of this section is to

M. Witczak: Model. and Estim. Strat. for Fault Diagn. of Non-Linear Syst. LNCIS 354, pp. 133–183, 2007.
springerlink.com

show how to describe model uncertainty of a neural network using the statistical framework. Another objective is to propose algorithms that can be used for developing an optimal experimental design which makes it possible to obtain a neural network with a possibly small uncertainty. The final objective is to show how to use the obtained knowledge about model uncertainty for robust fault diagnosis. The approach presented in this section is based on a static neural network, i.e., the multi-layer perceptron (see Section 3.1.1). It should also be pointed out that the results described in this section were originally presented in [182, 188].

An approach that can utilise either a static or a dynamic model structure is described in Section 7.2. This strategy is based on a similar idea as that of Section 7.1, but instead of using a statistical description of model uncertainty a deterministic bounded-error [116, 170] approach is employed. In particular, one objective is to show how to describe model uncertainty of the so-called GMDH (Group Method of Data Handling) neural network. Another objective is to show how to use the obtained knowledge about model uncertainty for robust fault diagnosis. Finally, it should be pointed out that the presented results are based on [185].

## 7.1 Robust Fault Detection with the Multi-layer Perceptron

Let us consider a feed-forward neural network given by the following equation:

$$\boldsymbol{y}_{M,k} = \boldsymbol{P}^{(l)}\boldsymbol{g}\left(\boldsymbol{P}^{(n)}\boldsymbol{u}_k\right), \tag{7.1}$$

while $\boldsymbol{g}(\cdot) = [g_1(\cdot), \ldots, g_{n_h}(\cdot), 1]^T$, while $g_i(\cdot) = g(\cdot)$ is a non-linear differentiable activation function,

$$\boldsymbol{P}^{(l)} = \begin{bmatrix} \boldsymbol{p}^{(l)}(1)^T \\ \vdots \\ \boldsymbol{p}^{(l)}(m)^T \end{bmatrix}, \quad \boldsymbol{P}^{(n)} = \begin{bmatrix} \boldsymbol{p}^{(n)}(1)^T \\ \vdots \\ \boldsymbol{p}^{(n)}(n_h)^T \end{bmatrix}, \tag{7.2}$$

are matrices representing the parameters (weights) of the model, $n_h$ is the number of neurons in the hidden layer. Moreover, $\boldsymbol{u}_k \in \mathbb{R}^{r=n_r+1}$, $\boldsymbol{u}_k = [u_{1,k}, \ldots, u_{n_r,k}, 1]^T$, where $u_{i,k}$, $i = 1, \ldots, n_r$ are system inputs. For the sake of notational simplicity, let us define the following parameter vector:

$$\boldsymbol{p} = \left[\boldsymbol{p}^{(l)}(1)^T, \ldots, \boldsymbol{p}^{(l)}(m)^T, \boldsymbol{p}^{(n)}(1)^T, \ldots, \boldsymbol{p}^{(n)}(n_h)^T\right]^T,$$

where $n_p = m(n_h + 1) + n_h(n_r + 1)$. Consequently, the equation (7.1) can be written in a more compact form:

$$\boldsymbol{y}_{M,k} = \boldsymbol{h}\left(\boldsymbol{p}, \boldsymbol{u}_k\right), \tag{7.3}$$

where $\boldsymbol{h}\left(\cdot\right)$ is a non-linear function representing the structure of a neural-network.

Let us assume that the system output satisfies the following equality:

$$\boldsymbol{y}_k = \boldsymbol{y}_{M,k} + \boldsymbol{v}_k = \boldsymbol{h}\left(\boldsymbol{p}, \boldsymbol{u}_k\right) + \boldsymbol{v}_k, \tag{7.4}$$

where the noise $\boldsymbol{v}$ is zero-mean, Gaussian and uncorrelated in $k$, i.e., its statistics are

$$\mathcal{E}(\boldsymbol{v}_k) = \boldsymbol{0}, \quad \mathcal{E}(\boldsymbol{v}_i \boldsymbol{v}_k^T) = \delta_{i,k}\boldsymbol{C}, \tag{7.5}$$

where $\boldsymbol{C} \in \mathbb{R}^{m\times m}$ is a known positive-definite matrix of the form $\boldsymbol{C} = \sigma^2 \boldsymbol{I}_m$, while $\sigma^2$ and $\delta_{i,k}$ stand for the variance and Kronecker's delta symbol, respectively. Under such an assumption, the theory of experimental design [7, 170] can be exploited to develop a suitable training data set that allows obtaining a neural network with a considerably smaller uncertainty than those designed without it. First, let us define the so-called Fisher information matrix (see also Section 5.1) that constitutes a measure of parametric uncertainty of (7.1):

$$\boldsymbol{P}^{-1} = \sum_{k=1}^{n_t} \boldsymbol{R}_k \boldsymbol{R}_k^T, \tag{7.6}$$

$$\boldsymbol{R}_k = \left(\frac{\partial \boldsymbol{h}\left(\boldsymbol{p}, \boldsymbol{u}_k\right)}{\partial \boldsymbol{p}}\right)^T_{\boldsymbol{p}=\hat{\boldsymbol{p}}}, \tag{7.7}$$

and

$$\frac{\partial \boldsymbol{h}\left(\boldsymbol{p}, \boldsymbol{u}_k\right)}{\partial \boldsymbol{p}} = \begin{bmatrix} \boldsymbol{g}\left(\boldsymbol{P}^{(n)}\boldsymbol{u}_k\right)^T & \boldsymbol{0}^T_{(m-1)(n_h+1)} & p_1^l(1)g'\left(\boldsymbol{u}_k^T p^n(1)\right)\boldsymbol{u}_k^T & \ldots & p_{n_h}^l(1)g'\left(\boldsymbol{u}_k^T p^n(n_h)\right)\boldsymbol{u}_k^T \\ \vdots & \vdots & \vdots & \vdots & \vdots \\ \boldsymbol{0}^T_{(m-1)(n_h+1)} & \boldsymbol{g}\left(\boldsymbol{P}^{(n)}\boldsymbol{u}_k\right)^T & p_1^l(m)g'\left(\boldsymbol{u}_k^T p^n(1)\right)\boldsymbol{u}_k^T & \ldots & p_{n_h}^l(m)g'\left(\boldsymbol{u}_k^T p^n(n_h)\right)\boldsymbol{u}_k^T \end{bmatrix}, \tag{7.8}$$

where $g'(t) = \dfrac{\mathrm{d}g(t)}{\mathrm{d}t}$, $\hat{\boldsymbol{p}}$ is a least-square estimate of $\boldsymbol{p}$. It is easy to observe that the FIM (7.6) depends on the experimental conditions $\xi = [\boldsymbol{u}_1, \ldots, \boldsymbol{u}_{n_t}]$. Thus, optimal experimental conditions can be found by choosing $\boldsymbol{u}_k$, $k = 1, \ldots, n_t$, so as to minimise some scalar function of (7.6). As was mentioned in Section 5.1, such a function can be defined in several different ways [61, 167]:

- D-optimality criterion:

$$\Phi(\xi) = \det \boldsymbol{P}, \tag{7.9}$$

- E-optimality criterion ($\lambda_{\max}(\cdot)$ stands for the maximum eigenvalue of its argument):

$$\Phi(\xi) = \lambda_{\max}\left(\boldsymbol{P}\right); \tag{7.10}$$

- A-optimality criterion:

$$\Phi(\xi) = \text{trace}\, \boldsymbol{P}; \tag{7.11}$$

- The G-optimality criterion:

$$\Phi(\xi) = \max_{\boldsymbol{u}_k \in \mathbb{U}} \phi(\xi, \boldsymbol{u}_k), \tag{7.12}$$

  and $\mathbb{U}$ stands for a set of admissible $\boldsymbol{u}_k$ that can be used for a system being considered (the design space), and

$$\phi(\xi, \boldsymbol{u}_k) = \text{trace}\left(\boldsymbol{R}_k^T \boldsymbol{P} \boldsymbol{R}_k\right) = \sum_{i=1}^{m} \boldsymbol{r}_{i,k} \boldsymbol{P} \boldsymbol{r}_{i,k}^T, \tag{7.13}$$

  while $\boldsymbol{r}_{i,k}$ stands for the $i$th row of $\boldsymbol{R}_k^T$;
- Q-optimality criterion:

$$\Phi(\xi) = \text{trace}\left(\boldsymbol{P}_Q^{-1} \boldsymbol{P}\right), \tag{7.14}$$

  where $\boldsymbol{P}_Q^{-1} = \int \boldsymbol{R}(\boldsymbol{u})\boldsymbol{R}(\boldsymbol{u})^T \mathrm{d}Q(\boldsymbol{u})$, $\boldsymbol{R}(\boldsymbol{u}) = \left(\frac{\partial \boldsymbol{h}(\boldsymbol{p},\boldsymbol{u})}{\partial \boldsymbol{p}}\right)^T_{\boldsymbol{p}=\hat{\boldsymbol{p}}}$, and $Q$ stands for the so-called environmental probability, which gives independent input vectors in the actual environment where a trained network is to be exploited [61].

As has already been mentioned, a valuable property of the FIM is that its inverse constitutes an approximation of the covariance matrix for $\hat{\boldsymbol{p}}$ [66], i.e., it is a lower bound of this covariance matrix that is established by the so-called Cramér-Rao inequality [66]:

$$\text{cov}(\hat{\boldsymbol{p}}) \succeq \boldsymbol{P}. \tag{7.15}$$

Thus, a D-optimum design minimises the volume of the confidence ellipsoid approximating the feasible parameter set of (7.1) (see, e.g., [7][Section 6.2] for further explanations). An E-optimum design minimises the length of the largest axis of the same ellipsoid. An A-optimum design suppresses the average variance of parameter estimates. A G-optimum design minimises the variance of the estimated response of (7.1). Finally, a Q-optimum design minimises the expectation of the generalisation error $\mathcal{E}(\varepsilon_{\text{gen}})$ defined by [61]:

$$\varepsilon_{\text{gen}} = \int \|\boldsymbol{h}(\boldsymbol{p}, \boldsymbol{u}) - \boldsymbol{h}(\hat{\boldsymbol{p}}, \boldsymbol{u})\|^2 \mathrm{d}Q(\boldsymbol{u}). \tag{7.16}$$

Among the above-listed optimality criteria, the D-optimality criterion, due to its simple updating formula (that is to be discussed in Section 7.1.2), has been employed by many authors in the development of computer algorithms for calculating optimal experimental design. Another important property is that D-optimum design is invariant to non-degenerate linear transformation of the model. This property is to be exploited and suitably discussed in Section 7.1.2. It is also important to underline that, from the practical point of view, D-optimum designs often perform well according to other criteria (see [7] and the references therein for more details). For further explanations regarding D-optimality criteria, the reader is referred to the excellent textbooks [7, 50, 170, 167]. Since the research results presented in this section are motivated by fault diagnosis applications of neural networks, the main objective is to use such a design criterion which makes it possible to obtain accurate bounds of the system output (cf. Fig. 7.1). Indeed, it is rather vain to assume that it is possible to develop a neural network with an arbitrarily small uncertainty, i.e., to obtain a perfect model of the system. A more realistic task is to design a model that will provide a reliable knowledge about the bounds of the system output that reflect the expected behaviuor of the system. As wasindicated in Section 2.3, this is especially important from the point of view of robust fault diagnosis. The design methodology of such robust techniques relies on the idea that fault diagnosis and control schemes should perform reliably for any system behaviour that is consistent with output bounds. This is in contradiction with the conventional approaches, where fault diagnosis and control schemes are designed to be optimal for one single model. The bounds presented in Fig. 7.1 can be described as follows:

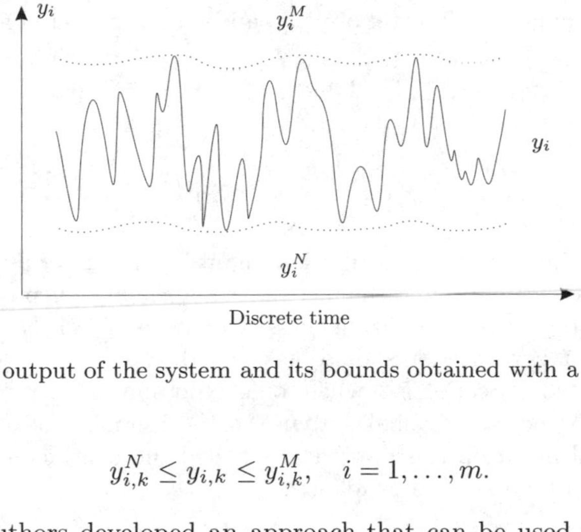


**Fig. 7.1.** $i$th output of the system and its bounds obtained with a neural network

$$y_{i,k}^N \leq y_{i,k} \leq y_{i,k}^M, \quad i = 1, \ldots, m. \tag{7.17}$$

In [32], the authors developed an approach that can be used for determining (7.17) (that forms the $100(1-\alpha)$ confidence interval of $y_{i,k}$) for single output ($m = 1$) neural networks. In [182], the approach of [32] was extended to multi-output models. If the neural model gives a good prediction of the actual system behaviuor, then $\hat{\boldsymbol{p}}$ is close to the optimal parameter vector and the following first-order Taylor expansion of (7.4) can be exploited:

$$\boldsymbol{y}_k \approx \hat{\boldsymbol{y}}_k + \boldsymbol{R}_k^T(\boldsymbol{p} - \hat{\boldsymbol{p}}) + \boldsymbol{v}_k, \quad \hat{\boldsymbol{y}}_k = \boldsymbol{h}(\hat{\boldsymbol{p}}, \boldsymbol{u}_k). \tag{7.18}$$

Thus, assuming that $\mathcal{E}(\hat{\boldsymbol{p}}) = \boldsymbol{p}$, we get

$$\mathcal{E}(\boldsymbol{y}_k - \hat{\boldsymbol{y}}_k) \approx \boldsymbol{R}_k^T(\boldsymbol{p} - \mathcal{E}(\hat{\boldsymbol{p}})) + \mathcal{E}(\boldsymbol{v}_k) \approx \boldsymbol{0}. \tag{7.19}$$

Using a similar approach, the covariance matrix is given by

$$\begin{aligned}\mathrm{cov}(\boldsymbol{y}_k - \hat{\boldsymbol{y}}_k) =& \mathcal{E}\left((\boldsymbol{y}_k - \hat{\boldsymbol{y}}_k)(\boldsymbol{y}_k - \hat{\boldsymbol{y}}_k)^T\right) \approx \\ &\approx \boldsymbol{R}_k^T \mathcal{E}\left((\boldsymbol{p} - \hat{\boldsymbol{p}})(\boldsymbol{p} - \hat{\boldsymbol{p}})^T\right) \boldsymbol{R}_k + \sigma^2 \boldsymbol{I}_m. \end{aligned} \tag{7.20}$$

Using the classic results regarding $\mathcal{E}\left((\boldsymbol{p} - \hat{\boldsymbol{p}})(\boldsymbol{p} - \hat{\boldsymbol{p}})^T\right)$ [7, 167, 170], i.e.,

$$\mathcal{E}\left((\boldsymbol{p} - \hat{\boldsymbol{p}})(\boldsymbol{p} - \hat{\boldsymbol{p}})^T\right) = \sigma^2 \boldsymbol{P}, \tag{7.21}$$

the equation (7.20) can be expressed as

$$\mathrm{cov}(\boldsymbol{y}_k - \hat{\boldsymbol{y}}_k) \approx \sigma^2 \left(\boldsymbol{R}_k^T \boldsymbol{P} \boldsymbol{R}_k + \boldsymbol{I}_m\right). \tag{7.22}$$

Subsequently, using (7.22), the standard deviation of $y_{i,k} - \hat{y}_{i,k}$ is given by

$$\sigma_{y_{i,k} - \hat{y}_{i,k}} = \sigma \left(1 + \boldsymbol{r}_{i,k} \boldsymbol{P} \boldsymbol{r}_{i,k}^T\right)^{1/2}, \quad i = 1, \ldots, m. \tag{7.23}$$

Using (7.23) and the result of [32], it can be shown that $y_{i,k}^N$ and $y_{i,k}^M$ (that form the $100(1-\alpha)$ confidence interval of $y_{i,k}$) can be approximated as follows:

$$y_{i,k}^N = \hat{y}_{i,k} - t_{n_t - n_p}^{\alpha/2} \hat{\sigma} \left(1 + \boldsymbol{r}_{i,k} \boldsymbol{P} \boldsymbol{r}_{i,k}^T\right)^{1/2}, \ i = 1, \ldots, m, \tag{7.24}$$

$$y_{i,k}^M = \hat{y}_{i,k} + t_{n_t - n_p}^{\alpha/2} \hat{\sigma} \left(1 + \boldsymbol{r}_{i,k} \boldsymbol{P} \boldsymbol{r}_{i,k}^T\right)^{1/2}, \ i = 1, \ldots, m, \tag{7.25}$$

where $t_{n_t - n_p}^{\alpha/2}$ is the t-Student distribution quantile, and $\hat{\sigma}$ is the standard deviation estimate. Bearing in mind the fact that the primary purpose is to develop reliable bounds of the system output, it is clear from (7.17), (7.24), and (7.25) that the G-optimality criterion should be selected.

As was indicated in Section 5.2, when some experiments are repeated then the number $n_e$ of distinct $\boldsymbol{u}_k$s is smaller than the total number of observations $n_t$. The design resulting from this approach is called the continuous experimental design and it can be described as follows:

$$\xi = \left\{ \begin{matrix} \boldsymbol{u}_1 & \boldsymbol{u}_2 & \ldots & \boldsymbol{u}_{n_e} \\ \mu_1 & \mu_2 & \ldots & \mu_{n_e} \end{matrix} \right\}, \tag{7.26}$$

where $\boldsymbol{u}_k$s are said to be the *support points*, and $\mu_1, \ldots, \mu_{n_e}, \mu_k \in [0,1]$ are called their weights, which satisfy $\sum_{k=1}^{n_e} \mu_k = 1$. Thus, when the design (7.26) is optimal (with respect to one of the above-defined criteria), then the support points can also be called *optimal inputs*. Thus, the Fisher information matrix can now be defined as follows:

$$\boldsymbol{P}^{-1} = \sum_{k=1}^{n_e} \mu_k \boldsymbol{R}_k \boldsymbol{R}_k^T. \tag{7.27}$$

The fundamental property of continuous experimental design is the fact that the optimum designs resulting from the D-optimality and G-optimality criteria are the same (the Kiefer-Wolfowitz equivalence theorem [7, 170, 167]). Another reason for using D-optimum design is the fact that it is probably the most popular criterion. Indeed, most of the algorithms presented in the literature are developed for D-optimum design. Bearing in mind all of the above-mentioned circumstances, the subsequent part of this section is devoted to D-optimum experimental design. The next section shows an illustrative example whose results clearly show profits that can be gained while applying D-OED to neural networks.

### 7.1.1 Illustrative Example

Let us consider a neuron model with the logistic activation function [182]:

$$y_{M,k} = \frac{p_1}{1 + e^{-p_2 u_k - p_3}}. \tag{7.28}$$

It is obvious that the continuous experimental design for the model (7.28) should have at least three different support points ($n_p = 3$ for (7.28)). For a three-point design, the determinant of the FIM (7.27) is

$$\det \boldsymbol{P}^{-1} = \frac{p_1^4}{p_2^2}\mu_1\mu_2\mu_3 e^{2x_1}e^{2x_2}e^{2x_3} \cdot \frac{((e^{x_2}-e^{x_1})x_3 + (e^{x_3}-e^{x_2})x_1 + (e^{x_1}-e^{x_3})x_2)^2}{(e^{x_1}+1)^4(e^{x_2}+1)^4(e^{x_3}+1)^4}, \tag{7.29}$$

where $x_i = p_2 u_i + p_3$. Bearing in mind the fact that the minimisation of (7.9) is equivalent to the maximisation of (7.29), a numerical solution regarding the D-optimum continuous experimental design can be written as

$$\xi = \left\{ \begin{matrix} \boldsymbol{u}_1 & \boldsymbol{u}_2 & \boldsymbol{u}_3 \\ \mu_1 & \mu_2 & \mu_3 \end{matrix} \right\} = \left\{ \begin{matrix} \left(\frac{1.041-p_3}{p_2}, 1\right) & \left(\frac{-1.041-p_3}{p_2}, 1\right) & \left(\frac{x_3-p_3}{p_2}, 1\right) \\ \frac{1}{3} & \frac{1}{3} & \frac{1}{3} \end{matrix} \right\}, \tag{7.30}$$

whereas $x_3$ is an arbitrary constant satisfying $x_3 \geq \zeta$, $\zeta \approx 12$. In order to check if the design (7.30) is really D-optimum, the Kiefer-Wolfowitz equivalence theorem [7, 170] can be employed. In the light of this theorem, the design (7.30) is D-optimum when

$$\phi(\xi, \boldsymbol{u}_k) = \operatorname{trace}\left(\boldsymbol{R}_k^T \boldsymbol{P} \boldsymbol{R}_k\right) \leq n_p, \tag{7.31}$$

where the equality holds for measurements described by (7.30). It can be seen from Fig. 7.2 that the design (7.30) satisfies (7.31). This figure justifies also the role of the constant $\zeta$, which is a lower bound of $x_3$ in the third support point

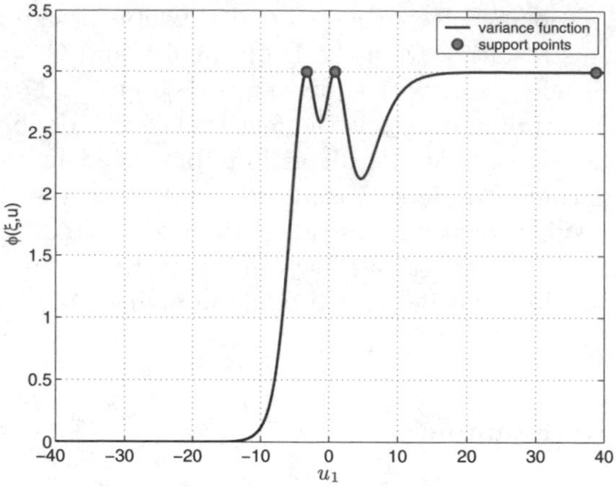


**Fig. 7.2.** Variance function for (7.30) and $x_3 = 20$

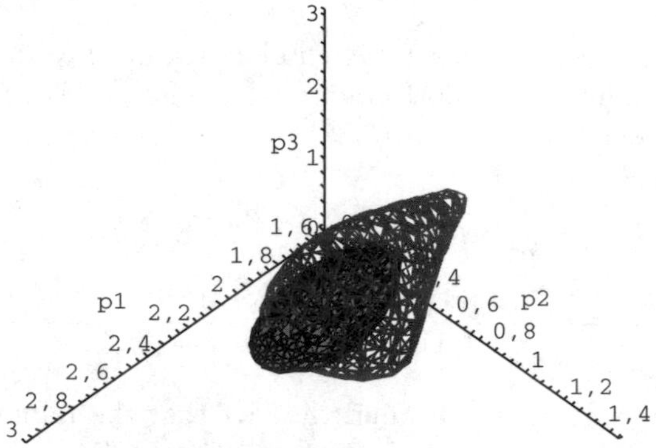


**Fig. 7.3.** Feasible parameter set obtained for (7.30) (smaller) and for a set of randomly generated points

of (7.30). Indeed, it can be observed that the design (7.30) is D-optimum (the variance function is $n_p = 3$) when the third support point is larger than some constant value, which is equivalent to $x_3 \geq \zeta$.

In order to justify the effectiveness of (7.30), let us assume that the nominal parameter vector is $\boldsymbol{p} = [2, 0.5, 0.6]^T$. It is also assumed that $n_t = 9$. This means that each of the measurements consistent with (7.30) should be repeated 3 times. For the purpose of comparison, a set of $n_t$ points was generated according to the uniform distribution $\mathcal{U}(-4, 40)$. It should also be pointed out that $v$ was generated according to $\mathcal{N}(0, 0.1^2)$. Fig. 7.3 presents feasible parameter sets obtained with the strategies considered. These sets are defined according to the following formula [170]:

$$\mathbb{P} = \left\{ \boldsymbol{p} \in \mathbb{R}^{n_p} \,\middle|\, \sum_{k=1}^{n_t} (y_k - f(\boldsymbol{p}, \boldsymbol{u}_k))^2 \leq \sigma^2 \chi^2_{\alpha,n_t} \right\}, \tag{7.32}$$

where $\chi^2_{\alpha,n_t}$ is the Chi-square distribution quantile. From Fig. 7.3, it is clear that the application of D-OED results in a model with a considerably smaller uncertainty than the one designed without it. These results also imply that the system bounds (7.17) will be more accurate.

### 7.1.2 Algorithms and Properties of D-OED for Neural Networks

#### Regularity of the FIM

The Fisher information matrix $\boldsymbol{P}^{-1}$ of (7.1) may be singular for some parameter configurations, and in such cases it is impossible to obtain its inverse $\boldsymbol{P}$ necessary to calculate (7.17), as well as to utilise specialised algorithms for obtaining D-optimum experimental design [7, 170]. Fukumizu [60] established the conditions under which $\boldsymbol{P}^{-1}$ is singular. These conditions can be formulated as follows:

**Theorem 7.1.** *[60] The Fisher information matrix $\boldsymbol{P}^{-1}$ of (7.1) is singular iff at least one of the following conditions holds true:*

1. *There exists $j$ such that $[p^{(n)}_{j,1}, \ldots, p^{(n)}_{j,n_r}]^T = \boldsymbol{0}$;*
2. *There exists $j$ such that $[p^{(l)}_{1,j}, \ldots, p^{(l)}_{m,j}]^T = \boldsymbol{0}$;*
3. *There exist different $j_1$ and $j_2$ such that $\boldsymbol{p}^{(n)}(j_1) = \pm\boldsymbol{p}^{(n)}(j_2)$.*

A direct consequence of the above theorem is that a network with singular $\boldsymbol{P}^{-1}$ can be reduced to one with positive definite $\boldsymbol{P}^{-1}$ by removing redundant hidden neurons. Based on this property, it is possible to develop a procedure that can be used for removing the redundant neurons without performing the retraining of a network [59].

If the conditions of Theorem 7.1 indicate that $\boldsymbol{P}^{-1}$ is not singular, then the strategy of collecting measurements according to the theory of D-optimum experimental design (the maximisation of the determinant of $\boldsymbol{P}^{-1}$) guarantees that the Fisher information matrix is positive definite. This permits approximating an exact feasible parameter set (7.32) with an ellipsoid (cf. Fig. 7.3 to see the similarity to an ellipsoid). Unfortunately, the conditions of Theorem 7.1 have strictly theoretical meaning as in most practical situations the FIM would be close to singular but not singular in an exact sense. This makes the process of eliminating redundant hidden neurons far more difficult, and there is no really efficient algorithm that could be employed to settle this problem. Indeed, the approach presented in [61] is merely sub-optimal. On the other hand, if such an algorithm does not give satisfactory results, i.e., the FIM is still close to the singular matrix, then the FIM should be regularised in the following way [7, p. 110]:

$$\boldsymbol{P}_{\kappa}^{-1} = \boldsymbol{P}^{-1} + \kappa \boldsymbol{I}, \tag{7.33}$$

for $\kappa > 0$ small but large enough to permit the inversion of $\boldsymbol{P}_{\kappa}^{-1}$.

### Relation between non-linear parameters and D-OED

Dependence on parameters that enter non-linearly ((7.30) depends on $p_2$ and $p_3$ but does not depend on $p_1$) into the model is an unappealing characteristic of non-linear optimum experimental design. As has already been mentioned, there is a number of works dealing with D-OED for neural networks but none of them has exploited this important property. In [182], it was shown that experimental design for a general structure (7.1) is independent of parameters that enter linearly into (7.1). Indeed, it can be shown that (7.8) can be transformed into an equivalent form:

$$\frac{\partial \boldsymbol{h}(\boldsymbol{p}, \boldsymbol{u}_k)}{\partial \boldsymbol{p}} = \boldsymbol{L}\left(\boldsymbol{P}^{(n)}, \boldsymbol{u}_k\right) \boldsymbol{Z}\left(\boldsymbol{P}^{(l)}\right), \tag{7.34}$$

and

$$\boldsymbol{L}\left(\boldsymbol{P}^{(n)}, \boldsymbol{u}_k\right) = \\ = \begin{bmatrix} \boldsymbol{g}\left(\boldsymbol{P}^{(n)}\boldsymbol{u}_k\right)^T & \boldsymbol{0}_{(m-1)(n_h+1)} & \left(\boldsymbol{g}'\left(\boldsymbol{P}^{(n)}\boldsymbol{u}_k\right) \otimes \boldsymbol{u}_k\right)^T \\ \vdots & \vdots & \vdots \\ \boldsymbol{0}_{(m-1)(n_h+1)} & \boldsymbol{g}\left(\boldsymbol{P}^{(n)}\boldsymbol{u}_k\right)^T & \left(\boldsymbol{g}'\left(\boldsymbol{P}^{(n)}\boldsymbol{u}_k\right) \otimes \boldsymbol{u}_k\right)^T \end{bmatrix}, \tag{7.35}$$

$$\boldsymbol{Z}\left(\boldsymbol{P}^{(l)}\right) = \begin{bmatrix} \boldsymbol{1}_{n_h+1} & \dots & \boldsymbol{x}_{(m-1)(n_h+1)} \\ \boldsymbol{x}_{(m-1)(n_h+1)} & \dots & \boldsymbol{1}_{n_h+1} \\ p_1^{(l)}(1)\boldsymbol{1}_{n_r+1} & \dots & p_1^{(l)}(m)\boldsymbol{1}_{n_r+1} \\ \vdots & \vdots & \vdots \\ p_{n_h+1}^{(l)}(1)\boldsymbol{1}_{n_r+1} & \dots & p_{n_h+1}^{(l)}(m)\boldsymbol{1}_{n_r+1} \end{bmatrix}, \tag{7.36}$$

where $\otimes$ denotes the Kronecker product, $\boldsymbol{x}_t$ stands for an arbitrary $t$-dimensional vector, and

$$\boldsymbol{g}'(t) = [g'(t_1), \dots, g'(t_{n_h})]^T. \tag{7.37}$$

Thus, $\boldsymbol{R}_k$ can be written in the following form:

$$\boldsymbol{R}_k = \boldsymbol{P}_1 \boldsymbol{R}_{1,k}, \tag{7.38}$$

where

$$\boldsymbol{P}_1 = \left(\boldsymbol{Z}\left(\boldsymbol{P}^{(l)}\right)\right)^T_{\boldsymbol{p}=\hat{\boldsymbol{p}}} \tag{7.39}$$

and

$$\boldsymbol{R}_{1,k} = \left(\boldsymbol{L}\left(\boldsymbol{P}^{(n)}, \boldsymbol{u}_k\right)\right)^T_{\boldsymbol{p}=\hat{\boldsymbol{p}}}. \tag{7.40}$$

The Fisher information matrix is now given by

$$\boldsymbol{P}^{-1} = \sum_{k=1}^{n_t} \boldsymbol{R}_k \boldsymbol{R}_k^T = \boldsymbol{P}_1 \left[\sum_{k=1}^{n_t} \boldsymbol{R}_{1,k} \boldsymbol{R}_{1,k}^T\right] \boldsymbol{P}_1^T. \tag{7.41}$$

Thus, the determinant of $\boldsymbol{P}^{-1}$ is given by

$$\det\left(\boldsymbol{P}^{-1}\right) = \det\left(\boldsymbol{P}_1\right)^2 \det\left(\sum_{k=1}^{n_t} \boldsymbol{R}_{1,k}\boldsymbol{R}_{1,k}^T\right). \tag{7.42}$$

From (7.42) it is clear that the process of minimising the determinant of $\boldsymbol{P}^{-1}$ with respect to $\boldsymbol{u}_k$s is independent of the linear parameters $\boldsymbol{p}^l$. This means that at least a rough estimate of $\boldsymbol{P}^{(n)}$ is required to solve the experimental design problem. Such estimates can be obtained with any training method for feed-forward neural networks [73]. A particularly interesting approach was recently developed in [46]. The authors proposed a novel method of backpropagating the desired response through the layers of the MLP in such a way as to minimise the mean-square error. Thus, the obtained solution may constitute a good starting point for experimental design.

Indeed, it is rather vain to expect that it is possible to obtain a design that is to be appropriate for all networks of a given structure. It is very easy to imagine two neural networks of the same structure that may represent two completely different systems. If some rough estimates are given, then specialised algorithms for D-optimum experimental design can be applied [7, 170].

### $\sigma$-equivalence theorem for the D-OED

Undoubtedly, the most popular activation functions $g(\cdot)$ that are commonly employed for designing neural networks are $g_\sigma(t) = \frac{1}{1+\exp(-t)}$ and $g_{\text{tg}}(t) = \tanh(t)$. It is well known that these functions are very similar, and this similarity is expressed by the following relationship:

$$g_\sigma(t) = \frac{1}{2} + \frac{1}{2} g_{\text{tg}}\left(\frac{1}{2}t\right). \tag{7.43}$$

Thus, the problem is to show how to use a D-optimum design obtained for a network with the activation functions $g_\sigma(\cdot)$ to obtain a D-optimum design for a network with the activation functions $g_{\text{tg}}(\cdot)$. In this work, the above problem is solved as follows:

**Theorem 7.2.** *Let*

$$\xi_\sigma = \begin{Bmatrix} \boldsymbol{u}_1 \dots \boldsymbol{u}_{n_e} \\ \mu_1 \dots \mu_{n_e} \end{Bmatrix} \tag{7.44}$$

*denote a D-optimum design for the network*

$$\boldsymbol{y}_{M,k} = \boldsymbol{P}^{(l)} \boldsymbol{g}_{\text{tg}}\left(\boldsymbol{P}^{(n)}\boldsymbol{u}_k\right). \tag{7.45}$$

*Then the design (7.44) is D-optimum for the following network:*

$$\boldsymbol{y}_{M,k} = \boldsymbol{P}_\sigma^{(l)} \boldsymbol{g}_\sigma\left(\boldsymbol{P}^{(n)}\boldsymbol{u}_k^\sigma\right), \tag{7.46}$$

*where $\boldsymbol{u}_k^\sigma = 2\boldsymbol{u}_k$, and $\boldsymbol{P}_\sigma^{(l)}$ is an arbitrary (non-zero) matrix.*

*Proof.* It is straightforward to observe that

$$\boldsymbol{P}_{\sigma}^{(l)}\boldsymbol{g}_{\sigma}\left(2\boldsymbol{P}^{(n)}\boldsymbol{u}_k\right)=\boldsymbol{P}_{\mathrm{tg}}^{(l)}\boldsymbol{g}_{\mathrm{tg}}\left(\boldsymbol{P}^{(n)}\boldsymbol{u}_k\right), \tag{7.47}$$

where

$$\boldsymbol{P}_{\mathrm{tg}}^{(l)}=\frac{1}{2}\boldsymbol{P}_{\sigma}^{(l)}+\left[\boldsymbol{0}_{m\times n_h},\frac{1}{2}\boldsymbol{P}_{\sigma}^{(l)}\boldsymbol{1}_{n_h+1}\right]. \tag{7.48}$$

Thus, using (7.42), the determinant of the FIM for (7.45) is

$$\det\left(\boldsymbol{P}^{-1}\right)=\det\left(\left(\boldsymbol{Z}\left(\boldsymbol{P}^{(l)}\right)\right)^{T}_{\boldsymbol{p}=\hat{\boldsymbol{p}}}\right)^{2}\det\left(\sum_{k=1}^{n_t}\boldsymbol{R}_{1,k}\boldsymbol{R}_{1,k}^{T}\right) \tag{7.49}$$

while using (7.42), (7.47), and (7.48), the determinant of the Fisher information matrix for (7.46) is

$$\det\left(\boldsymbol{P}^{-1}\right)=\det\left(\left(\boldsymbol{Z}\left(\boldsymbol{P}_{\mathrm{tg}}^{(l)}\right)\right)^{T}_{\boldsymbol{p}=\hat{\boldsymbol{p}}}\right)^{2}\det\left(\sum_{k=1}^{n_t}\boldsymbol{R}_{1,k}\boldsymbol{R}_{1,k}^{T}\right), \tag{7.50}$$

and $\boldsymbol{R}_{1,k}$ in (7.49) and (7.50) is calculated by substituting

$$\boldsymbol{L}\left(\boldsymbol{P}^{(n)},\boldsymbol{u}_k\right)=$$

$$=\begin{bmatrix}\boldsymbol{g}_{\mathrm{tg}}\left(\boldsymbol{P}^{(n)}\boldsymbol{u}_k\right)^{T} & \boldsymbol{0}_{(m-1)(n_h+1)} & \left(\boldsymbol{g}_{\mathrm{tg}}'\left(\boldsymbol{P}^{(n)}\boldsymbol{u}_k\right)\otimes\boldsymbol{u}_k\right)^{T}\\ \vdots & \vdots & \vdots\\ \boldsymbol{0}_{(m-1)(n_h+1)} & \boldsymbol{g}_{\mathrm{tg}}\left(\boldsymbol{P}^{(n)}\boldsymbol{u}_k\right)^{T} & \left(\boldsymbol{g}_{\mathrm{tg}}'\left(\boldsymbol{P}^{(n)}\boldsymbol{u}_k\right)\otimes\boldsymbol{u}_k\right)^{T}\end{bmatrix}, \tag{7.51}$$

and

$$\boldsymbol{g}_{\mathrm{tg}}'(t)=[g_{\mathrm{tg}}'(t_1),\ldots,g_{\mathrm{tg}}'(t_{n_h})]^{T}, \tag{7.52}$$

into (7.40).

From (7.49) and (7.50), it is clear that the D-optimum design obtained with either (7.49) or (7.50) is identical, which completes the proof.

Based on the above results, the following remark can be formulated:

*Remark 7.1.* Theorem 7.2 and the Kiefer-Wolfowitz theorem [7, 170] imply that the D-optimum design (7.44) for (7.45) is also G-optimum for this model structure and, hence, it is G-optimum for (7.46).

### Illustrative example

Let us reconsider the example presented in Section 7.1.1. The purpose of this example was to obtain D-optimum experimental design for the model (7.28). As a result, the design (7.30) was determined. The purpose of further deliberations is to apply the design (7.30) to the following model:

$$y_{M,k}=p_1\tanh(2(p_2u_k+p_3)), \tag{7.53}$$

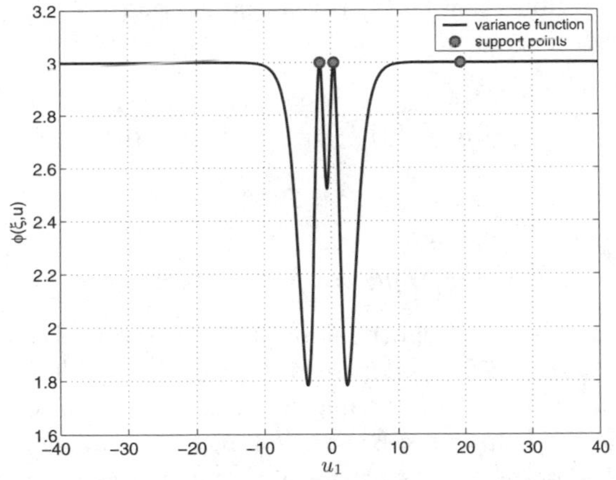


**Fig. 7.4.** Variance function for (7.30) ($x_3 = 20$) and model (7.53)

and to check if it is D-optimum for (7.53). Figure 7.4 presents the variance function (7.12) for the model (7.53). From this figure it is clear that the variance function satisfies the D-optimality condition (7.31).

### Wynn–Fedorov algorithm for the MLP

The preceding part of Section 7.1.2 presents important properties of D-OED for neural networks. In this section, these properties are exploited to develop an effective algorithm for calculating D-OED for neural networks. In a numerical example of Section 7.1.1, it is shown how to calculate D-OED for a neural network composed of one neuron only. In particular, the algorithm was reduced to a direct optimisation of the determinant of the FIM with respect to experimental conditions. This means that non-linear programming techniques have to be employed to settle this problem. Unfortunately, it should be strongly underlined that such an approach is impractical when larger neural networks are investigated. Fortunately, the Kiefer-Wolfowitz equivalence theorem [7, 170, 167] (see also (7.31)) provides some guidance useful in construction of a suitable numerical algorithm. The underlying reasoning boils down to a correction of a non-optimum design $\xi_k$ (obtained after $k$ iterations) by a convex combination with another design [50, p. 27], which hopefully improves the current solution, i.e.,

$$\xi_{k+1} = (1 - \alpha_k)\xi_k + \alpha_k \xi(\boldsymbol{u}_k) \tag{7.54}$$

for some convenient $0 < \alpha_k < 1$, where

$$\xi_k = \begin{Bmatrix} \boldsymbol{u}_1 \, \boldsymbol{u}_2 \dots \, \boldsymbol{u}_{n_e} \\ \mu_1 \, \mu_2 \, \dots \, \mu_{n_e} \end{Bmatrix}, \quad \xi(\boldsymbol{u}_k) = \begin{Bmatrix} \boldsymbol{u}_k \\ 1 \end{Bmatrix}, \tag{7.55}$$

while the convex combination (7.54) is realised as follows:
If $\boldsymbol{u}_k \neq \boldsymbol{u}_i$, $i = 1, \ldots, n_e$, then

$$\xi_{k+1} = \left\{ \begin{matrix} \boldsymbol{u}_1 & \boldsymbol{u}_2 & \ldots & \boldsymbol{u}_{n_e} & \boldsymbol{u}_k \\ (1-\alpha_k)\mu_1 & (1-\alpha_k)\mu_2 & \ldots & (1-\alpha_k)\mu_{n_e} & \alpha_k \end{matrix} \right\}, \tag{7.56}$$

else

$$\xi_{k+1} = \left\{ \begin{matrix} \boldsymbol{u}_1 & \boldsymbol{u}_2 & \ldots & \boldsymbol{u}_i & \ldots & \boldsymbol{u}_{n_e} \\ (1-\alpha_k)\mu_1 & (1-\alpha_k)\mu_2 & \ldots & (1-\alpha_k)\mu_i + \alpha_k & \ldots & (1-\alpha_k)\mu_{n_e} \end{matrix} \right\}.$$

In this way, the experimental effort related to $\xi_k$ is reduced and measurements corresponding to $\xi(\boldsymbol{u}_k)$ are favored instead. Hence, the problem is to select $\xi(\boldsymbol{u}_k)$ so as to get a better value of the optimality criterion. A solution to this problem can be found with the help of the Kiefer-Wolfowitz equivalence theorem [7, 170, 167]. Indeed, the support points of the optimum design $\xi^*$ coincide with the maxima of the variance function $\phi(\xi^*, \boldsymbol{u}_k)$ (see Section 7.1.1 for an illustrative example). Thus, by the addition of $\xi(\boldsymbol{u}_k)$ for which the maximum of $\phi(\xi_k, \boldsymbol{u}_k)$ is attained, an improvement in the current design can be expected (see [170, 167] for more details).

The above-outlined approach forms the base of the celebrated Wynn-Fedorov algorithm [7, 170]. In this section, the results developed in Section (7.1.2) and the one of Theorem 7.1 are utilised to adapt the Wynn-Fedorov algorithm in order to develop D-OED for neural networks.

First, let us start with a slight modification of the Wynn-Fedorov algorithm that boils down to reducing the necessity of using the linear parameters of (7.1) in the computational procedure.

Since, according to (7.38),

$$\boldsymbol{R}_k = \boldsymbol{P}_1 \boldsymbol{R}_{1,k} , \tag{7.57}$$

then (7.12) can be written as follows:

$$\Phi(\xi) = \max_{\boldsymbol{u}_k \in \mathbb{U}} \operatorname{trace} \left( \boldsymbol{R}_{1,k}^T \boldsymbol{P}_1^T \boldsymbol{P} \boldsymbol{P}_1 \boldsymbol{R}_{1,k} \right). \tag{7.58}$$

Using (7.41) and the notation of continuous experimental design, the matrix $\boldsymbol{P}$ can be expressed as follows:

$$\boldsymbol{P} = \left( \boldsymbol{P}_1^T \right)^{-1} \left[ \sum_{k=1}^{n_e} \mu_k \boldsymbol{R}_{1,k} \boldsymbol{R}_{1,k}^T \right]^{-1} \boldsymbol{P}_1^{-1}. \tag{7.59}$$

It is easy to observe that if Condition 2 of Theorem 7.1 is not satisfied, then it is possible to compute the inverse of $\boldsymbol{P}_1$ (cf. (7.39)). Similarly, if both Conditions 1 and 3 are not satisfied, then the matrix

$$\boldsymbol{P}_2 = \left[ \sum_{k=1}^{n_e} \mu_k \boldsymbol{R}_{1,k} \boldsymbol{R}_{1,k}^T \right]^{-1} \tag{7.60}$$

in (7.59) can be calculated. Now, (7.58) can be expressed in the following form:

$$\Phi(\xi) = \max_{\boldsymbol{u}_k \in \mathbb{U}} \phi(\xi, \boldsymbol{u}_k), \tag{7.61}$$

where

$$\phi(\xi, \boldsymbol{u}_k) = \text{trace}\left(\boldsymbol{R}_{1,k}^T \boldsymbol{P}_2 \boldsymbol{R}_{1,k}\right). \tag{7.62}$$

Note that the computation of (7.61) does not require any knowledge about the parameter matrix $\boldsymbol{P}^{(l)}$, which enters linearly into (7.1). Another advantage is that it is not necessary to form the matrix $\boldsymbol{P}_1 \in \mathbb{R}^{n_p \times n_p}$, which then has to be multiplied by $\boldsymbol{R}_{1,k}$ to form (7.57). This implies a reduction of the computational burden.

The equation (7.61) can be perceived as the main step of the Wynn-Fedorov algorithm, which can now be described as follows:

Step 0: Obtain an initial estimate of $\boldsymbol{P}^{(n)}$, i.e., the parameter matrix that enters non-linearly into (7.1), with any method for training the MLP [73]. Set $k = 1$, choose a non-degenerate ($\det(\boldsymbol{P}^{-1}) \neq 0$, it is satisfied when no conditions of Theorem 7.1 are fulfilled) design $\xi_k$, set the maximum number of iterations $n_{\max}$.

Step 1: Calculate

$$\boldsymbol{u}_k = \arg \max_{\boldsymbol{u}_k \in \mathbb{U}} \text{trace}\left(\boldsymbol{R}_{1,k}^T \boldsymbol{P}_2 \boldsymbol{R}_{1,k}\right). \tag{7.63}$$

Step 2: If $\phi(\xi_k, \boldsymbol{u}_k)/n_p < 1 + \epsilon$, where $\epsilon > 0$, is sufficiently small, then STOP, else go to *Step 3*.

Step 3: Calculate a weight associated with a new support point $\boldsymbol{u}_k$ according to

$$\alpha_k = \arg \max_{0<\alpha<1} \det\left((1-\alpha)\boldsymbol{P}_2 + \alpha \boldsymbol{R}_{1,k} \boldsymbol{R}_{1,k}^T\right), \tag{7.64}$$

which for single-output systems ($m = 1$) is given by

$$\alpha_k = \frac{\phi(\xi_k, \boldsymbol{u}_k) - n_p}{(\phi(\xi_k, \boldsymbol{u}_k) - 1)n_p}, \tag{7.65}$$

and go to *Step 4*.

Step 4: Obtain a new design $\xi_{k+1}$ being a convex combination [50, p. 27] of the form

$$\xi_{k+1} = (1 - \alpha_k)\xi_k + \alpha_k \xi(\boldsymbol{u}_k). \tag{7.66}$$

If $k = n_{\max}$, then STOP, else set $k = k + 1$ and go to *Step 1*.

*Step 1* is a crucial step of the presented algorithm. Indeed, the first problem is the fact that the calculation of (7.60) involves matrix inversions. Since the dimension of this matrix equals $n_p$, then, even for simple networks, the number of parameters is a dozen or so. Subsequently, it is shown that effective recursive formulae can be established for calculating $\boldsymbol{P}_2^k$, i.e., the matrix $\boldsymbol{P}_2$ in the $k$th iteration of the Wynn-Fedorov algorithm. It can be seen from the inverse of (7.60) and (7.56) that

$$\left(\boldsymbol{P}_2^{k+1}\right)^{-1} = (1-\alpha_k)\left(\boldsymbol{P}_2^{k}\right)^{-1} + \alpha_k \boldsymbol{R}_{1,k}\boldsymbol{R}_{1,k}^T. \tag{7.67}$$

Using the matrix inversion lemma and (7.67), the following recursive relation can be established:

$$\begin{aligned}\boldsymbol{P}_2^{k+1} =& \frac{1}{1-\alpha_k}\cdot \\ &\cdot \left[\boldsymbol{P}_2^k - \boldsymbol{P}_2^k\boldsymbol{R}_{1,k}\left[\frac{1-\alpha_k}{\alpha_k}\boldsymbol{I}_m + \boldsymbol{R}_{1,k}^T\boldsymbol{P}_2^k\boldsymbol{R}_{1,k}\right]^{-1}\boldsymbol{R}_{1,k}^T\boldsymbol{P}_2^k\right].\end{aligned} \tag{7.68}$$

Note that the calculation of (7.68) requires an inversion of an $m$-dimensional matrix instead of an $n_p$-dimensional one.

The second problem concerning *Step 1* is the fact that the variance function (7.63) is multi-modal and hence conventional optimisation routines cannot be applied to settle (7.61). For further explanations concerning the problem (7.61), the reader is referred to [17]. Based on numerous computer experiments, it has been found that the extremely simple Adaptive Random Search (ARS) algorithm [170] is especially well suited for the purpose of optimising (7.61), although other techniques such as evolutionary algorithms [96] can successfully be applied as well.

It is important to note that the above algorithm makes use of information about the gradient of the performance index only, and the rule (7.64) results in the steepest-descent algorithm. As a result, the convergence rate of the algorithm is comparable with its gradient counterparts from mathematical non-linear programming. This implies a significant decrease in the performance index in the first few iterations, but then a serious moderation of the convergence rate occurs as the optimum is approached. There are some second-order counterparts of the algorithm being considered, but they require a significantly higher implementation complexity. However, it should be pointed out that they may improve the design weight rather than the support points, and in this context the features of the presented algorithm are satisfactory. Indeed, many computer experiments show that the most significant support points are found in just several iterations.

### Numerical example

The problem is to approximate the function

$$y_k = \exp(-\sin(u_k)) + v_k,$$

where $v \sim \mathcal{N}(0, 0.02^2)$, $u_k \in [0.1, 10]$, with a neural network containing $n_h = 4$ hidden neurons with hyperbolic tangent activation functions. Thus, the number of parameters to be estimated is $n_p = 13$. In the preliminary experiment $u_k, k = 1, \ldots, n_t = 15$ were obtained in such a way as to equally divide the

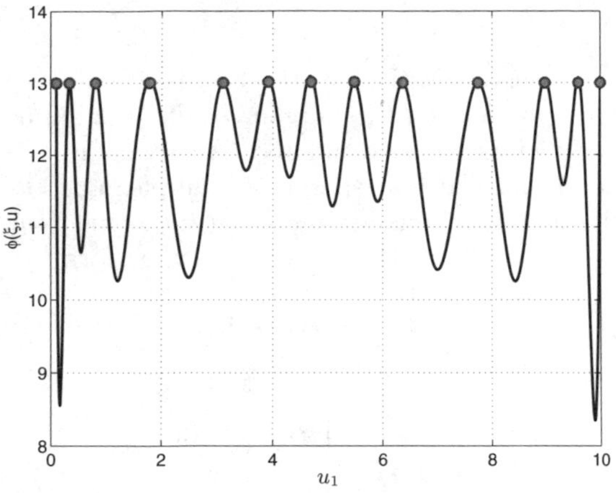


**Fig. 7.5.** Variance function and the corresponding support points

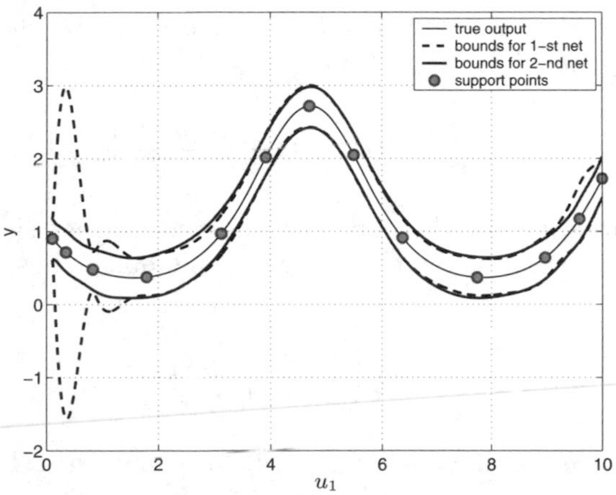


**Fig. 7.6.** Output and its bounds

design space $\mathbb{U} \in [0.1, 10]$. Then the Levenberg-Marquardt algorithm [170] was employed for parameter estimation. Based on the obtained parameter estimates, the Wynn-Fedorov algorithm was utilised to obtain D-OED, and then the parameter estimation process was repeated once again. Figure 7.5 shows the variance function and D-optimum inputs (support points). Note that the number of support points is $n_p$ while $\mu_k = 1/13,\ k = 1, \ldots, 13$. Based on the obtained design, $n_t = 13$ measurements were taken, each corresponding to the subsequent support points. Figure 7.6 presents the output bounds (7.17) for the network obtained with the application of OED (the 2nd net) and the one obtained without it (the 1st net), while the true output represents the shape of the approximated function. It can be observed that the use of OED results in a network with a significantly smaller uncertainty than the one designed without it.

## Special case

A special case that deserves particular attention is when the design consists of $n_e = n_p/m$ support points, i.e., the number of distinct support points equals the number of parameters to be estimated. An example of such a design is presented in Section 7.1.2. The analysis of this example indicates that the weights associated with the support points are the same and equal $1/n_p$. This can easily be explained by transforming (7.41) into the following form:

$$\boldsymbol{P}^{-1} = \boldsymbol{R}^T \boldsymbol{W} \boldsymbol{R}, \tag{7.69}$$

where

$$\boldsymbol{R} = \begin{bmatrix} \boldsymbol{R}_1^T \\ \vdots \\ \boldsymbol{R}_{n_e}^T \end{bmatrix}, \tag{7.70}$$

and $\boldsymbol{W} = \operatorname{diag}(\mu_1 \mathbf{1}_m, \ldots, \mu_{n_e} \mathbf{1}_m)$.

It can be observed from (7.70) that $\boldsymbol{R}$ is a square matrix when $n_p = mn_e$. To achieve this, the number of support points should be

$$n_e = \frac{n_p}{m} = (n_h + 1) + \frac{n_h(n_r + 1)}{m} \tag{7.71}$$

As can be seen from (7.71), the number of hidden neurons should be suitably selected to guarantee that $n_e$ is a positive integer number. Thus, the determinant of (7.69) is

$$\det\left(\boldsymbol{P}^{-1}\right) = \left(\prod_{k=1}^{n_e} \mu_k\right)^m \det\left(\boldsymbol{R}\right)^2. \tag{7.72}$$

From (7.72), it is clear that $\mu_k$, $k = 1, \ldots, n_p/m$ maximising $\det\left(\boldsymbol{P}^{-1}\right)$ are the same and equal $m/n_p$.

Unfortunately, it is impossible to expect a priori how many support points should be used to form a D-optimum design for a given neural model. Indeed, the equation (7.71) indicates the minimum number of support points, while the maximum number can be determined with the help of Caratheodory's theorem [7, 170, 167] and is equal to $n_e = n_p(n_p + 1)/2 + 1$.

Now let us consider a single output neural model for which $n_e = n_p$. The parameter vector of (7.1) is estimated with the least-square method as follows:

$$\hat{\boldsymbol{p}} = \arg \min_{\boldsymbol{p} \in \mathbb{R}^{n_p}} \sum_{k=1}^{n_t} (y_k - h(\boldsymbol{p}, \boldsymbol{u}_k))^2. \tag{7.73}$$

Another appealing characteristic of the design being considered can be expressed by the following theorem, which is based on the results presented in [147]:

**Theorem 7.3.** *Assume that*

$$\xi = \left\{ \begin{matrix} \boldsymbol{u}_1 \dots \boldsymbol{u}_{n_p} \\ \frac{1}{n_p} \dots \frac{1}{n_p} \end{matrix} \right\},$$

*and the number of observations for different* $\boldsymbol{u}_k$*s is* $n_x = \frac{n_t}{n_p}$ *(it is assumed that it is a positive integer number). Assume also that for all* $\boldsymbol{\beta} \in \mathbb{R}^{n_p}$*, there exists* $\boldsymbol{p} \in \mathbb{R}^{n_p}$ *such that* $\boldsymbol{\beta} = [h(\boldsymbol{p}, \boldsymbol{u}_1), \dots, h(\boldsymbol{p}, \boldsymbol{u}_{n_p})]^T$ *and* $\boldsymbol{p}$ *does not satisfy the conditions of Theorem 7.1. Then the cost function of (7.73) has a unique global minimiser* $\hat{\boldsymbol{p}}$ *and no other global minimisers.*

*Proof.* See the proof of Theorem 1 in [147].

Since the weights associated with support points are the same, then it is natural to assume that the number of observations for different $\boldsymbol{u}_k$s is the same. Theorem 7.3 can be relatively easily interpreted because the optimisation problem (7.73) can be expressed (under the assumptions of Theorem 7.3) as

$$\min_{\boldsymbol{p} \in \mathbb{R}^{n_p}} \sum_{k=1}^{n_p} (\bar{y}_k - h(\boldsymbol{p}, \boldsymbol{u}_k))^2 = 0, \tag{7.74}$$

and

$$\bar{y}_k = \frac{1}{n_x} \sum_{i=1}^{n_x} y_k^i, \tag{7.75}$$

where $y_k^i$ stands for the $i$th observation under $\boldsymbol{u}_k$. Thus, the solution $\hat{\boldsymbol{p}}$ of (7.74) should satisfy

$$\bar{y}_k - h(\hat{\boldsymbol{p}}, \boldsymbol{u}_k) = 0, \quad k = 1, \dots, n_p. \tag{7.76}$$

Indeed, the fact that $\hat{\boldsymbol{p}}$ does not satisfy the conditions of Theorem 7.1 implies that (7.1) is uniquely determined by its input-output map, up to a finite group of symmetries (permutations of hidden neurons and changing the sign of all weights associated with a particular hidden neuron) [164]. This means that $\hat{\boldsymbol{p}}$ is a unique solution of (7.76).

### Towards robustness – sequential design

As was shown in Section 7.1.2, the unappealing characteristic of experimental design for the MLP is the fact that the FIM depends on the non-linear parameters $\boldsymbol{P}^{(n)}$ only. It is obvious that the true value of $\boldsymbol{P}^{(n)}$ is unknown and hence its estimate should be utilised instead. As was mentioned in Section 7.1.2, if some rough estimates are given, i.e., they can be obtained with any training method for feed-forward neural networks [73], then the so-called *sequential design* [167, 170] can be applied. Such a strategy is usually applied off-line, i.e., the first step is parameter estimation while the second one is to use some specialised algorithms, e.g., the Wynn-Federov algorithm detailed in Section 7.1.2, to obtain a design of the form (7.26). In spite of the simplicity of such a sequential approach, some non-trivial problems arise which can be described as follows:

- Determination of the number of stages of experimentation-estimation required to attain the prescribed accuracy;
- Dependence of the final design upon the initial parameter estimates;
- Unique parametrisation of (7.1). This is the necessary condition to ensure the convergence of the sequential algorithm;
- Management of data collected in the consecutive experiments in order to guarantee the convergence of $\hat{\boldsymbol{p}}$ to the true value of parameters $\boldsymbol{p}$.

Some existing results being partial solutions to the first two questions can be found in [49, 55, 170]. Sussman [164] proved that, under some conditions, a network of the structure (7.1) with the hyperbolic tangent activation function is uniquely determined by its input-output map, up to a finite group of symmetries (permutations of hidden neurons and changing the sign of all weights associated with a particular hidden neuron). Fukumizu [60] extended the results of [164] to the structure (7.1) with the logistic activation function. The solution to the last problem seems to be well developed and can be formulated as follows [170]: in order to guarantee the convergence of $\hat{\boldsymbol{p}}$ to $\boldsymbol{p}$, the estimation of $\hat{\boldsymbol{p}}^k$ (the estimate of $\boldsymbol{p}$ in the $k$th iteration of the sequential algorithm) should make use of all previous observations collected during the preceding iterations of the sequential algorithm. Fukumizu [61] employed this strategy for OED for the MLP. The routine employed in [61] adds one single measurement to the measurement set collected in the preceding iterations of the algorithm. This new support point is obtained in such a way as to obtain an optimum design for the new parameter estimate. In Section 7.1.3, this idea is exploited in designing a new sequential algorithm that can be used for both training and data development for the MLP. Another approach [170] is to obtain a design for a new parameter estimate in a classic way, e.g., with the Wynn-Fedorov algorithm, while parameter estimation should make use of all the previous observations that where collected during the preceding iterations of the sequential algorithm. This strategy is employed in the numerical example presented in the subsequent section.

### Numerical example

Let us reconsider the example presented in Section 7.1.1. It is assumed that an initial parameter estimate is $\hat{\boldsymbol{p}} = [1.8, 0.45, 0.54]^T$. The sequential algorithm utilises (7.30) to obtain OED in the consecutive iterations of the algorithm. The measurements $y$ were generated by disturbing the data obtained with (7.28) by the normally distributed random noise $\mathcal{N}(0, 0.1^2)$. The Levenberg-Marquardt algorithm [170] was employed for parameter estimation. In order to show the reliability of the sequential algorithm, consisting of 250 cycles of estimation and experimentation, it was repeated 100 times. This means that in each cycle, $9 \times 250$ measurements were collected, i.e., in each iteration the measurements were repeated three times for each support point of (7.30).

Figure 7.7 shows an average norm of the parameter estimation error $\|\boldsymbol{p} - \hat{\boldsymbol{p}}\|_2$ in the consecutive iterations of the sequential algorithm. From this result it can be seen that the parameter estimate converges (on average) to the true

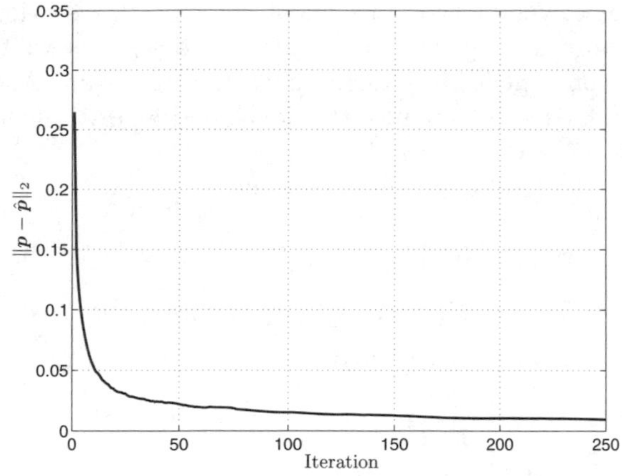


**Fig. 7.7.** Average norm of the parameter estimation error in the consecutive iterations of the sequential algorithm

parameter vector $\boldsymbol{p}$. This implies that the designed experiment tends to the optimal experiment for $\boldsymbol{p}$.

### 7.1.3 Locally D-Optimum On-Line Sequential Design

An alternative sequential design approach, different than that proposed in Section 7.1.2, presented in [146] consists in modifying a recursive parameter estimation technique in such a way as to develop an algorithm that can be used for both obtaining a one-step-ahead D-optimum input and estimating the parameters of the model. In general, however, this policy is not globally optimal and that is why it should be called the locally D-optimum sequential design. Unfortunately, the algorithm developed by [146] belongs to the class of the so-called bounded-error estimation techniques [170], and it can be used for parameter estimation of linear-in-parameter models only. Thus, it seems especially attractive to adapt this approach to least-square parameter estimation and experimental design for neural networks. The purpose of the subsequent section is to propose an algorithm that tackles such a challenging problem. It should be strongly underlined that the algorithm developed in the subsequent part of this work can be applied to single-output systems ($m = 1$) only. However, in the final part of this section a possible extension for multi-output systems is discussed.

#### Development of the algorithm

For the purpose of further deliberations, a recursive technique that permits parameter estimation in a least-square sense is required. Thus, the recursive least-square [73, 170] algorithm seems to be a good choice. Indeed, some authors (see [73] and the references therein) recommend to use this technique for parameter estimation of neural networks.

One of the objectives of the subsequent part of this section is to modify RLS for the purpose of sequential experimental design as well as to perform a comprehensive convergence analysis of such an algorithm. Another objective is to propose a minor modification of the developed algorithm that significantly increases its performance.

Let us consider the RLS algorithm given by

$$\hat{\boldsymbol{p}}_{k+1} = \hat{\boldsymbol{p}}_k + \boldsymbol{k}_{k+1}\varepsilon_{k+1}, \tag{7.77}$$

$$\boldsymbol{k}_{k+1} = \boldsymbol{P}_k\boldsymbol{r}_{k+1}\left(\lambda_k + \boldsymbol{r}_{k+1}^T\boldsymbol{P}_k\boldsymbol{r}_{k+1}\right)^{-1}, \tag{7.78}$$

$$\varepsilon_{k+1} = y_{k+1} - h(\hat{\boldsymbol{p}}_k, \boldsymbol{u}_{k+1}), \tag{7.79}$$

$$\boldsymbol{P}_{k+1} = \frac{1}{\lambda_k}\left[\boldsymbol{I}_{n_p} - \boldsymbol{k}_{k+1}\boldsymbol{r}_{k+1}^T\right]\boldsymbol{P}_k, \tag{7.80}$$

where $\boldsymbol{r}_{k+1} = \left.\frac{\partial h(\boldsymbol{p},\boldsymbol{u}_{k+1})}{\partial \boldsymbol{p}}\right|_{\boldsymbol{p}=\hat{\boldsymbol{p}}_k}$, $\lambda_k$ stands for the so-called forgetting factor. As has already been mentioned, the problem consists in obtaining a sequence of $\boldsymbol{u}_k$s (note that $\boldsymbol{r}_k$ depends on $\boldsymbol{u}_k$) minimising the determinant of $\boldsymbol{P}_{k+1}$. From (7.80), it can be seen that

$$\det\left(\boldsymbol{P}_{k+1}\right) = \det\left(\frac{1}{\lambda_k}\boldsymbol{P}_k\right)\det\left(\boldsymbol{I}_{n_p} - \frac{\boldsymbol{P}_k\boldsymbol{r}_{k+1}\boldsymbol{r}_{k+1}^T}{\lambda_k + \boldsymbol{r}_{k+1}^T\boldsymbol{P}_k\boldsymbol{r}_{k+1}}\right). \tag{7.81}$$

Bearing in mind the fact that $\det(\boldsymbol{I}+\boldsymbol{a}\boldsymbol{b}^T) = 1 + \boldsymbol{b}^T\boldsymbol{a}$, the equation (7.81) can be written as

$$\det\left(\boldsymbol{P}_{k+1}\right) = \frac{\lambda_k^{-n_p}}{1+\lambda_k^{-1}\boldsymbol{r}_{k+1}^T\boldsymbol{P}_k\boldsymbol{r}_{k+1}}\det\left(\boldsymbol{P}_k\right). \tag{7.82}$$

Thus $\boldsymbol{u}_{k+1}$ minimising $\det\left(\boldsymbol{P}_{k+1}\right)$ should be given as

$$\boldsymbol{u}_{k+1}^* = \arg\max_{\boldsymbol{u}_{k+1}\in\mathbb{U}} \boldsymbol{r}_{k+1}^T\boldsymbol{P}_k\boldsymbol{r}_{k+1}, \tag{7.83}$$

and $\mathbb{U}$ stands for a set of admissible $\boldsymbol{u}_{k+1}$ that can be applied to the system being considered (the design space). As can be seen from (7.82), $\lambda_k$ can also be used for controlling $\det\left(\boldsymbol{P}_{k+1}\right)$. In order to make the subsequent presentation more intelligible, let us provide a detailed description of the proposed algorithm:

*Step 0:* Set $k=0$, obtain an initial parameter estimate $\hat{\boldsymbol{p}}_0$ with any method for training the MLP [73], set $\boldsymbol{P}_0 = \varrho\boldsymbol{I}$, where $\varrho$ is a sufficiently large positive constant;

*Step 1:* Obtain $\boldsymbol{u}_{k+1} = \boldsymbol{u}_{k+1}^*$ according to (7.83), and measure the system output $y_{k+1}$ for such an input;

*Step 2:* Calculate $\hat{\boldsymbol{p}}_{k+1}$ according to (7.77)–(7.80). If $k = n_t$ ($n_t$ being a predefined number of input-output measurements), then STOP, else set $k = k+1$ and go to *Step 1*.

It should be clearly pointed out that *Step 1* involves the global optimisation task of $\boldsymbol{r}_{k+1}^T \boldsymbol{P}_k \boldsymbol{r}_{k+1}$. Similarly as in Section 7.1.2, this problem can be solved with the ARS algorithm [170]. Another important conclusion that can be drawn while analysing (7.77)–(7.80) is that the matrix $\boldsymbol{P}_k^{-1}$ cannot be decomposed in the way described by (7.41) since the matrix $P_{1,k}$ varies in $k$, as it depends on the $k$th parameter estimate $\hat{\boldsymbol{p}}_k$.

### Development of the algorithm

Since the rule of obtaining locally D-optimum $\boldsymbol{u}_{k+1}$ is given, let us consider the convergence of such an algorithm. For that purpose, the Lyapunov approach is employed. The approach presented here is similar to that described in [18], which was employed for the convergence analysis of Kalman filter-based estimators (see also Sections 4.1 and 4.2).

The approach has proven to be useful for unknown input observers as well [179, 183]. The main objective is to determine conditions under which the sequence $\{V_k\}_{k=1}^{\infty}$, defined by the Lyapunov candidate function

$$V_{k+1} = \boldsymbol{e}_{k+1}^T \boldsymbol{P}_{k+1}^{-1} \boldsymbol{e}_{k+1} \tag{7.84}$$

is decreasing, where

$$\boldsymbol{e}_{k+1} = \boldsymbol{p} - \hat{\boldsymbol{p}}_{k+1}. \tag{7.85}$$

First, let us derive an alternative form of (7.78) and $\boldsymbol{P}_{k+1}^{-1}$. Substituting (7.78) into (7.80) and then using the matrix inversion lemma, $\boldsymbol{P}_{k+1}^{-1}$ can be written as

$$\boldsymbol{P}_{k+1}^{-1} = \lambda_k \boldsymbol{P}_k^{-1} + \boldsymbol{r}_{k+1} \boldsymbol{r}_{k+1}^T. \tag{7.86}$$

Similarly, applying the matrix inversion lemma to (7.78),

$$\boldsymbol{k}_{k+1} = \boldsymbol{P}_{k+1} \boldsymbol{r}_{k+1}. \tag{7.87}$$

the application of the classic approximation yields

$$\varepsilon_{k+1} \approx \boldsymbol{r}_{k+1}^T \boldsymbol{e}_k. \tag{7.88}$$

In order to avoid the above approximation, it is proposed to introduce an unknown scalar $\alpha_{k+1}$. This makes it possible to establish the following exact equality:

$$\varepsilon_{k+1} \alpha_{k+1} = \boldsymbol{r}_{k+1}^T \boldsymbol{e}_k. \tag{7.89}$$

Substituting (7.77) into (7.85) leads to

$$\boldsymbol{e}_{k+1} = \boldsymbol{e}_k - \boldsymbol{k}_{k+1} \varepsilon_{k+1}. \tag{7.90}$$

Using (7.87), (7.90) and (7.84), the Lyapunov candidate function becomes

$$V_{k+1} = \boldsymbol{e}_k^T \boldsymbol{P}_{k+1}^{-1} \boldsymbol{e}_k - 2\varepsilon_{k+1} \boldsymbol{r}_{k+1}^T \boldsymbol{e}_k + \varepsilon_{k+1}^2 \boldsymbol{r}_{k+1}^T \boldsymbol{k}_{k+1}. \tag{7.91}$$

Substituting (7.89), (7.86) and then (7.78) into (7.91) gives

$$V_{k+1} = \lambda_k \boldsymbol{e}_k^T \boldsymbol{P}_k^{-1} \boldsymbol{e}_k + \varepsilon_{k+1}^2 \left( \alpha_{k+1}^2 - 2\alpha_{k+1} + \frac{\boldsymbol{r}_{k+1}^T \boldsymbol{P}_k \boldsymbol{r}_{k+1}}{\lambda_k + \boldsymbol{r}_{k+1}^T \boldsymbol{P}_k \boldsymbol{r}_{k+1}} \right). \tag{7.92}$$

The sequence $\{V_k\}_{k=1}^{\infty}$ is decreasing when there exists a scalar $\zeta$, $0 < \zeta < 1$ such that

$$V_{k+1} - (1-\zeta)V_k \leq 0, \tag{7.93}$$

and hence

$$\begin{aligned} &(\lambda_k - 1 + \zeta)\boldsymbol{e}_k^T \boldsymbol{P}_k^{-1} \boldsymbol{e}_k + \\ &+ \varepsilon_{k+1}^2 \left( \alpha_{k+1}^2 - 2\alpha_{k+1} + \frac{\boldsymbol{r}_{k+1}^T \boldsymbol{P}_k \boldsymbol{r}_{k+1}}{\lambda_k + \boldsymbol{r}_{k+1}^T \boldsymbol{P}_k \boldsymbol{r}_{k+1}} \right) \leq 0. \end{aligned} \tag{7.94}$$

It is easy to see that (7.94) is equivalent to the following set of inequalities:

$$\lambda_k \leq 1 - \zeta, \tag{7.95}$$

$$\alpha_{k+1}^2 - 2\alpha_{k+1} + \frac{\boldsymbol{r}_{k+1}^T \boldsymbol{P}_k \boldsymbol{r}_{k+1}}{\lambda_k + \boldsymbol{r}_{k+1}^T \boldsymbol{P}_k \boldsymbol{r}_{k+1}} \leq 0. \tag{7.96}$$

Bearing in mind the fact that the left-hand-side of (7.96) is a quadratic function, it can be shown that (7.96) is equivalent to

$$1 - \sqrt{\Delta_{k+1}} \leq \alpha_{k+1} \leq 1 + \sqrt{\Delta_{k+1}}, \tag{7.97}$$

where

$$\Delta_{k+1} = \frac{1}{1 + \lambda_k^{-1} \boldsymbol{r}_{k+1}^T \boldsymbol{P}_k \boldsymbol{r}_{k+1}}. \tag{7.98}$$

Thus, $\Delta_{k+1}$ should be designed in such a way as to maximise the bounds (7.97) and hence enlarge the domain of attraction. Unfortunately, from (7.98) it is evident that selecting $\boldsymbol{u}_{k+1}$ according to (7.83) leads to a decrease in $\Delta_{k+1}$, i.e., it may lead to the divergence of the RLS algorithm. Another difficulty arises from the fact that $\lambda_k \leq 1$ (cf. (7.95)) and hence it cannot compensate for $\boldsymbol{r}_{k+1}^T \boldsymbol{P}_k \boldsymbol{r}_{k+1}$. A direct consequence of the above circumstances is that the algorithm (7.77)–(7.80) cannot be used for the purpose of sequential design, unless the parameter estimate $\hat{\boldsymbol{p}}_k$ is very close to its optimum.

### Improved version of the algorithm

In order to overcome the above difficulties, an alternative algorithm is proposed that is given by

$$\hat{\boldsymbol{p}}_{k+1} = \hat{\boldsymbol{p}}_k + \boldsymbol{k}_{k+1}\varepsilon_{k+1}, \tag{7.99}$$

$$\boldsymbol{k}_{k+1} = \boldsymbol{P}_{k+1/k}\boldsymbol{r}_{k+1}\left(\lambda_k + \boldsymbol{r}_{k+1}^T\boldsymbol{P}_{k+1/k}\boldsymbol{r}_{k+1}\right)^{-1}, \tag{7.100}$$

$$\varepsilon_{k+1} = y_{k+1} - h(\hat{\boldsymbol{p}}_k, \boldsymbol{u}_{k+1}), \tag{7.101}$$

$$\boldsymbol{P}_{k+1} = \frac{1}{\gamma_k}\left[\boldsymbol{I}_{n_p} - \boldsymbol{k}_{k+1}\boldsymbol{r}_{k+1}^T\right]\boldsymbol{P}_{k+1/k}, \tag{7.102}$$

$$\boldsymbol{P}_{k+1/k} = \boldsymbol{P}_k + \boldsymbol{Q}_k, \tag{7.103}$$

where $\boldsymbol{Q}_k$ stands for a symmetric positive-definite matrix and $\gamma_k$ denotes an arbitrary positive constant. Taking into account the result (7.95), the matrix $\boldsymbol{Q}_k$ is introduced to enhance the convergence of the algorithm, i.e., it should be designed in such a way as to prevent the divergence of the proposed approach. It should be stressed that the algorithm presented here is largely inspired by [18] except for the additional parameter $\gamma_k$.

For the algorithm (7.99)–(7.103), the determinant of $\boldsymbol{P}_{k+1}$ is

$$\det\left(\boldsymbol{P}_{k+1}\right) = \frac{\gamma_k^{-n_p}}{1 + \lambda_k^{-1}\boldsymbol{r}_{k+1}^T\boldsymbol{P}_{k+1/k}\boldsymbol{r}_{k+1}}\det\left(\boldsymbol{P}_{k+1/k}\right). \tag{7.104}$$

Thus, $\boldsymbol{u}_{k+1}$ minimising $\det\left(\boldsymbol{P}_{k+1}\right)$ should be given as

$$\boldsymbol{u}_{k+1}^* = \arg\max_{\boldsymbol{u}_{k+1}\in\mathbb{U}} \boldsymbol{r}_{k+1}^T\boldsymbol{P}_{k+1/k}\boldsymbol{r}_{k+1}, \tag{7.105}$$

while the convergence conditions equivalent to (7.95)–(7.96) are

$$\gamma_k(\boldsymbol{P}_k + \boldsymbol{Q}_k)^{-1} - (1-\zeta)\boldsymbol{P}_k^{-1} \preceq 0, \tag{7.106}$$

$$\alpha_{k+1}^2 - 2\alpha_{k+1} + \frac{\boldsymbol{r}_{k+1}^T\boldsymbol{P}_{k+1/k}\boldsymbol{r}_{k+1}}{\lambda_k + \boldsymbol{r}_{k+1}^T\boldsymbol{P}_{k+1/k}\boldsymbol{r}_{k+1}} \leq 0, \tag{7.107}$$

where

$$\boldsymbol{P}_{k+1}^{-1} = \gamma_k\boldsymbol{P}_{k+1/k}^{-1} + \frac{\gamma_k}{\lambda_k}\boldsymbol{r}_{k+1}\boldsymbol{r}_{k+1}^T. \tag{7.108}$$

Since (7.106) is equivalent to

$$(\gamma_k^{-1}\boldsymbol{P}_k + \gamma_k^{-1}\boldsymbol{Q}_k)^{-1} - \left(\gamma_k^{-1}\boldsymbol{P}_k + \frac{(1-(1-\zeta)\gamma_k^{-1})}{1-\zeta}\boldsymbol{P}_k\right)^{-1} \preceq 0, \tag{7.109}$$

then it is clear that the convergence conditions are

$$\left(\frac{\gamma_k}{1-\zeta} - 1\right)\boldsymbol{P}_k \preceq \boldsymbol{Q}_k, \tag{7.110}$$

and

$$1 - \sqrt{\Delta_{k+1}} \leq \alpha_{k+1} \leq 1 + \sqrt{\Delta_{k+1}}, \tag{7.111}$$

with

$$\Delta_{k+1} = \frac{1}{1 + \lambda_k^{-1} \boldsymbol{r}_{k+1}^T \boldsymbol{P}_{k+1/k} \boldsymbol{r}_{k+1}}. \tag{7.112}$$

Finally, to complete the convergence analysis, it is necessary to show that the FIM is bounded [18], i.e., that there exist positive scalars $\bar{\theta}$ and $\underline{\theta}$ such that

$$\mathbf{0} \prec \underline{\theta} \boldsymbol{I}_{n_p} \preceq \boldsymbol{P}_k^{-1} \preceq \bar{\theta} \boldsymbol{I}_{n_p}. \tag{7.113}$$

The relation (7.113) is associated with the so-called *persistency of excitation* (or local persistency of excitation since linearisation around a parameter estimate is employed) condition [18, 38, 67]. This condition is satisfied if the input signal $\boldsymbol{u}_k$ provides enough information to estimate $\boldsymbol{p}$. First, let us assume that the parameter estimate $\hat{\boldsymbol{p}}$ does not satisfy the conditions of Theorem 7.1. In the general framework proposed in Section 7.1.3, the input sequence is uniquely determined in *Step 1* by (7.105). Such an input selection guarantees that the determinant of $\boldsymbol{P}_k^{-1}$ is maximised in each iteration and, hence, the FIM is far from being a singular matrix. This implies that there exists a positive scalar $\underline{\theta} > 0$ such that $\boldsymbol{P}_k^{-1} \succeq \underline{\theta} \boldsymbol{I}_{n_p}$. On the other hand, since the input $\boldsymbol{u}_k$ is bounded, i.e., $\boldsymbol{u}_k \in \mathbb{U}$, then $\boldsymbol{P}_k^{-1} \preceq \bar{\theta} \boldsymbol{I}_{n_p}$ is also satisfied (cf. (7.108)). Following the results of [18], it can be deduced from (7.84), (7.93) and (7.113) that

$$0 \leq \underline{\theta} \|\boldsymbol{e}_k\|_2^2 \leq V_k \leq (1-\zeta)^k V_0 \leq \bar{\theta} \|\boldsymbol{e}_k\|_2^2 \tag{7.114}$$

and, therefore, $\|\boldsymbol{e}_k\|_2^2$ decreases. This implies that the developed algorithm is locally asymptotically convergent.

If the convergence of the algorithm is proven, then it is possible to provide some details regarding its implementation. In this work, the following setting of $\lambda_k$, $\gamma_k$, and $\boldsymbol{Q}_k$ is proposed:

$$\lambda_k = \beta_k \boldsymbol{r}_{k+1}^T \boldsymbol{P}_{k+1/k} \boldsymbol{r}_{k+1}, \tag{7.115}$$

$$\gamma_k = \delta_k \boldsymbol{r}_{k+1}^T \boldsymbol{P}_{k+1/k} \boldsymbol{r}_{k+1}, \tag{7.116}$$

$$\boldsymbol{Q}_k = \eta_k \varepsilon_k^2 \boldsymbol{I}_{n_p}, \tag{7.117}$$

where $\beta_k$, $\delta_k$, and $\eta_k$ are positive constants. Substituting (7.115) and (7.116) into (7.104) leads to

$$\det(\boldsymbol{P}_{k+1}) = \frac{\beta_k}{1+\beta_k} \frac{1}{\left(\delta_k \boldsymbol{r}_{k+1}^T \boldsymbol{P}_{k+1/k} \boldsymbol{r}_{k+1}\right)^{n_p}} \det\left(\boldsymbol{P}_{k+1/k}\right), \tag{7.118}$$

while inserting (7.115) into (113) gives

$$\Delta_{k+1} = \frac{\beta_k}{1+\beta_k}. \tag{7.119}$$

It can be seen from (7.119) that $\beta_k$ should be large when $\hat{\boldsymbol{p}}_k$ is far from its optimum. On the contrary, when $\hat{\boldsymbol{p}}_k$ is near to its optimum then $\beta_k$ should be small so as to minimise (7.118). Moreover, as the quality of estimates is improving, $\delta_k$ should converge to $\beta_k$ so that $\gamma_k \leq 1$. If $\gamma_k \leq 1$ then it can be seen from (7.110) and (7.117) that $\eta_k = 0$. Following the above rules, it can be observed that the algorithm (7.99)–(7.103) is reduced to the algorithm described by (7.77)–(7.80). This is very important from the point of view of (7.17), where $\boldsymbol{P}_k$ should be described in a proper way. Indeed, the modification performed in the proposed algorithm changes the primary role of $\boldsymbol{P}_k$.

If the convergence analysis is suitably performed and the necessary implementation details are provided, then it is possible to discuss a possible extension of the above algorithm for multi-output ($m > 1$) systems. Many researchers (see [73] and the references therein) employ the so-called extended Kalman filter for training multi-output feed-forward neural networks. Such an algorithm can be perceived as a multidimensional extension of the one presented in this section. Using ideas similar to those presented in this work and in [18], the authors of [19] developed convergence conditions for the extended Kalman filter (see also Sections 4.1 and 4.2). Thus, this result can also be used for convergence analysis of a multi-output counterpart of the proposed algorithm. This implies that the approach presented in this section can be extended to multi-output systems.

### Numerical example

Let us reconsider the example presented in Section 7.1.2. The preliminary steps were realised in the same way as in Section 7.1.2. Based on the obtained parameter estimates, the algorithm proposed in Section 7.1.3 was utilised to obtain the locally D-optimum input sequence and to estimate the parameters of the network. The settings of the algorithm were as follows:

$$\lambda_k = \max\left[0.6\boldsymbol{r}_{k+1}^T\boldsymbol{P}_{k+1/k}\boldsymbol{r}_{k+1}, 1\right], \tag{7.120}$$

$$\gamma_k = \max\left[0.005\boldsymbol{r}_{k+1}^T\boldsymbol{P}_{k+1/k}\boldsymbol{r}_{k+1}, 1\right], \tag{7.121}$$

$$\boldsymbol{Q}_k = \begin{cases} 0 & \lambda_k = 1 \\ 10\varepsilon_k^2\boldsymbol{I}_{n_p} & \text{otherwise} \end{cases}. \tag{7.122}$$

Figure 7.8 shows the obtained locally D-optimum input sequence, while Fig. 7.9 depicts the output bounds (7.17) for the network obtained with the proposed algorithm (the 2nd net) and the network obtained without it (the 1st net), whereas the true output represents the shape of the approximated function. From this it is clear that the application of the algorithm of Section 7.1.3 results in a network designed with a significantly smaller uncertainty than the one designed without it. Comparing Figs. 7.9 and 7.6, it can be observed that the output bounds obtained with the Wynn-Fedorov algorithm and those found with the approach of Section 7.1.3 are very similar. On the other hand, the support points determined by these techniques are different.

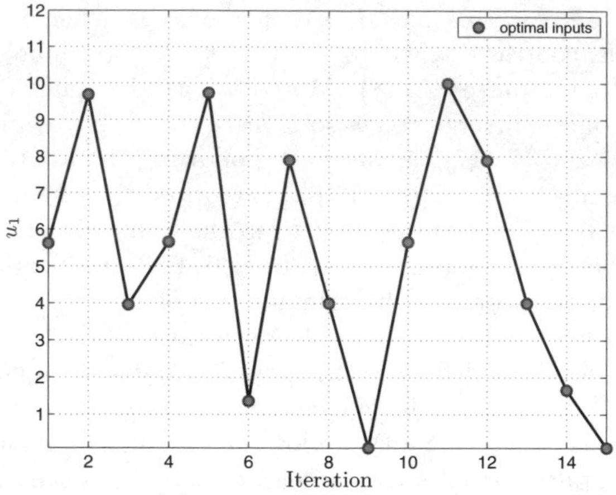


**Fig. 7.8.** Input sequence obtained with the proposed algorithm

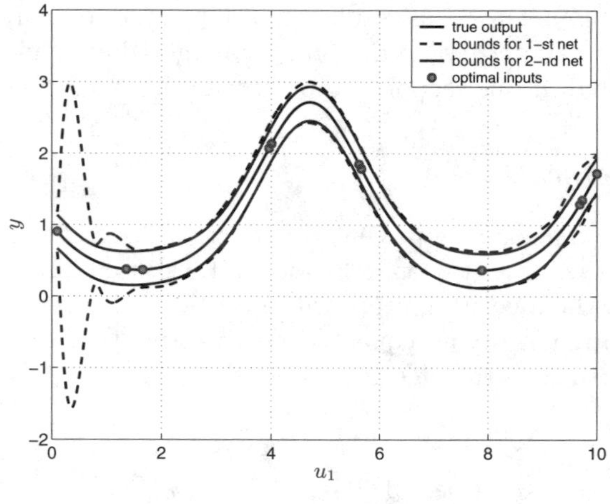


**Fig. 7.9.** Output of the system and its bounds obtained with a neural network

### 7.1.4 Industrial Application

The main objective of the subsequent part of this section is to develop a neural network that can be used for fault detection of the industrial valve actuator described in Section 6.4.2. The above task can be divided into the following steps:

*Step 1:* Training of a network based on the nominal data set;
*Step 2:* Design of the experiment with the Wynn-Fedorov algorithm described in Section 7.1.2 based on the network obtained in *Step 1*;
*Step 3:* Training of a network based on the data obtained with experimental design.

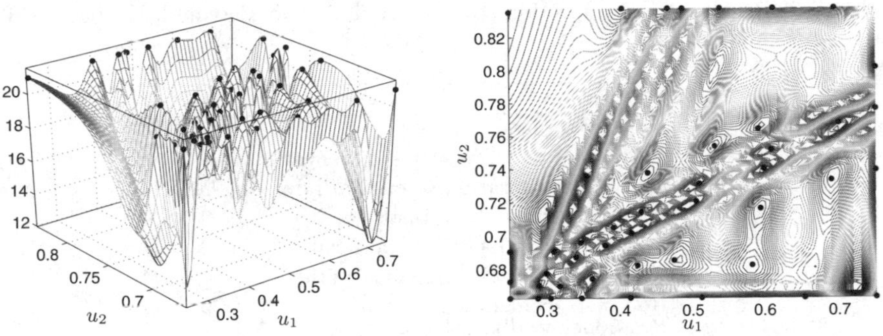


**Fig. 7.10.** Variance function and the corresponding support points

Based on the experience with the industrial valve actuator, it was observed that the following subset of the measured variables is sufficient for fault detection purposes: $\boldsymbol{u} = (CV, P1, 1)$, $y = F$.

In *Step 1*, a number of experiments (the training of a neural network with the Levenberg-Marquardt algorithm [170]) were performed in order to find a suitable number of hidden neurons $n_h$ (cf. (7.1)). For that purpose, $n_t = 100$ data points were generated for which inputs were uniformly spread within the design region $\mathbb{U}$, where $0.25 < u_1 < 0.75$ and $0.6625 < u_2 < 0.8375$. As a result, a neural model consisting of $n_h = 5$ hidden neurons was obtained. The main objective of *Step 2* was to utilise the above model and the Wynn-Fedorov algorithm in order to obtain D-optimum experimental conditions. First, an initial experiment was generated in such a way as to equally divide the design space of $\boldsymbol{u}$. Finally, the Wynn-Fedorov algorithm was applied. Figure 7.10 shows the support points ($n_e = 45$) and the variance function for the obtained D-optimum design. Based on the obtained continuous design, a set consisting of $n_t = 100$ points was found and used for data generation. The number of repetitions of each optimal input $\boldsymbol{u}_k$ was calculated by suitably rounding the numbers $\mu_k n_t$, $k = 1, \ldots, n_e$ [7, 170]. It should be strongly stressed that the data were collected in the steady-state of the valve because the utilised model (7.1) was static. Finally, the new data set was used for training the network with the Levenberg-Marquardt algorithm. As was mentioned at the beginning of this section, the research directions of the DAMADICS project were oriented towards fault diagnosis and, in particular, fault detection of the valve actuator. Under the assumption of a perfect mathematical description of the systems being considered, a perfect residual generation should provide a residual that is zero during the normal operation of the system and considerably different than zero otherwise. This means that the residual should ideally carry information regarding a fault only. Under such assumptions, faults can be easily detected. Unfortunately, this is impossible to attain in practice since residuals are normally uncertain, corrupted by noise, disturbances and modelling uncertainty. That is why in order to avoid false alarms it is necessary to assign a threshold to the residual that is significantly larger than zero. The most common approach is to use a fixed threshold [96, 179]. The

**Table 7.1.** Results of fault detection (D – detected, N – not detected, X – not specified for the benchmark)

| Fault | Description | S | M | B |
|---|---|---|---|---|
| $f_1$ | Valve clogging | D | D | D |
| $f_2$ | Valve plug or valve seat sedimentation | X | X | D |
| $f_7$ | Medium evaporation or critical flow | D | D | D |
| $f_8$ | Twisted servomotor's piston rod | N | N | N |
| $f_{10}$ | Servomotor's diaphragm perforation | D | D | D |
| $f_{12}$ | Electro-pneumatic transducer fault | X | X | D |
| $f_{13}$ | Rod displacement sensor fault | D | D | D |
| $f_{15}$ | Positioner feedback fault | X | X | D |
| $f_{16}$ | Positioner supply pressure drop | N | N | D |
| $f_{17}$ | Unexpected pressure change across valve | X | X | D |
| $f_{18}$ | Fully or partly opened bypass valves | D | D | D |
| $f_{19}$ | Flow rate sensor fault | D | D | D |

main difficulty with this kind of thresholds is the fact that they may cause many serious problems regarding false alarms as well as undetected faults. In other words, it is very difficult to fix such a threshold and there is no optimal solution that can be applied to settle such a task. Fortunately, using (7.17), (7.24) and (7.25) it is possible to develop an adaptive threshold that can be described as follows:

$$|z_{i,k}| \leq t^{\alpha/2}_{n_t-n_p}\hat{\sigma}\left(1+\boldsymbol{r}_{i,k}\boldsymbol{P}\boldsymbol{r}^T_{i,k}\right)^{1/2}, \quad i=1,\dots,m. \tag{7.123}$$

Consequently, the decision logic can be realised as follows:
*If the residual $\boldsymbol{z}_k$ satisfies (7.123), then there is no fault symptom, else (7.123) indicates that a fault symptom has occurred.*
The objective of the subsequent part of this section is to use the obtained network for fault detection, as well as to compare its performance with that of a network obtained for a nominal data set. Table 7.1 shows the results of fault detection for a set of faults being specified for the benchmark (the symbols S – Small, M – Medium, and B – Big denote the magnitude of the faults). All faults were generated with the same scenario, i.e., the first 200 samples correspond to the normal operating mode of the system while the remaining ones were generated under faulty conditions. Figure 7.11 presents the residual signal obtained with a network trained with the D-optimum data set as well as an adaptive threshold provided by this network (the 2nd network). This figure also presents an adaptive threshold provided by a network (the 1st network) trained with the data set generated by equally dividing the design space. It can be observed that the neural network obtained with the use of D-optimum experimental design makes it possible to obtain more accurate bounds than those obtained with a neural network trained otherwise. Indeed, as can be seen in Fig. 7.11b, the fault $f_1$ – small (which in the light of its nature is hard to detect) can be detected with the help of the 2nd network while it is impossible to detect it with the use of the 1st one. It should be strongly underlined that the situation is even worse when

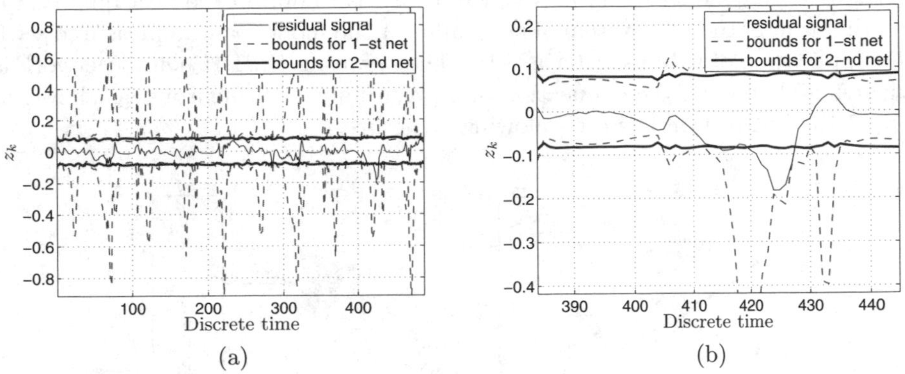


**Fig. 7.11.** Residual and adaptive thresholds for the fault $f_1$ – small (a) and its selected part (b)

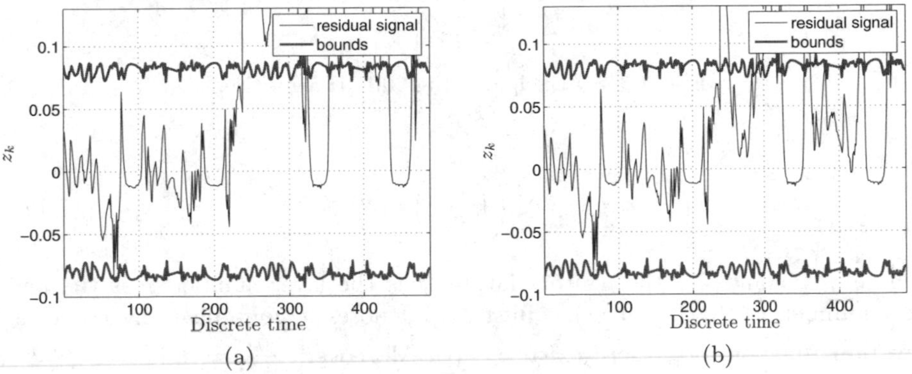


**Fig. 7.12.** Residual and adaptive thresholds for the fault $f_{18}$ – small (a) and $f_{19}$ – small (b), respectively

the 1st network is used for residual generation, i.e., in the presented example it was used for adaptive threshold generation only. As can be observed in Tab. 7.1, almost all the faults specified for the benchmark can be detected. The main reason why the faults $f_8$ and $f_{16}$ (small and medium) cannot be detected is because their effect is exactly at the same level as that of noise. However, it should be pointed out that this was the case for other techniques [128, 183] tested with the DAMADICS benchmark. Finally, Fig. 7.12 presents exemplary residuals for the faults $f_{18}$ – small and $f_{19}$ – small, respectively.

## 7.2 Robust Fault Detection with GMDH Neural Networks

The synthesis process of the GMDH neural network [185] is based on the iterative processing of a sequence of operations. This process leads to the evolution of the resulting model structure in such a way as to obtain the best quality approximation of the real system. The quality of the model can be measured with the application of various criteria [121]. The resulting GMDH neural network is constructed through the connection of a given number of neurons, as shown in Fig. 7.13. The neuron has the following structure:

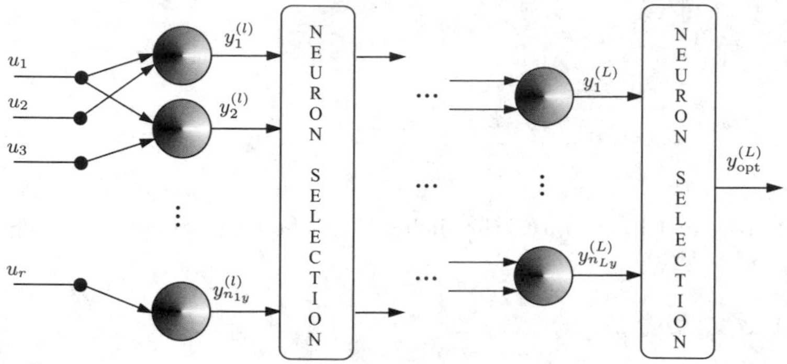


**Fig. 7.13.** Principle of the GMDH algorithm

$$y_{n,k}^{(l)} = \xi\left(\left(\boldsymbol{r}_{n,k}^{(l)}\right)^T \boldsymbol{p}_n^{(l)}\right), \tag{7.124}$$

where $y_{n,k}^{(l)}$ stands for the neuron output ($l$ is the layer number, $n$ is the neuron number in the $l$th layer), whilst $\xi(\cdot)$ denotes a non-linear invertible activation function, i.e., there exists $\xi^{-1}(\cdot)$. Moreover, $\boldsymbol{r}_{n,k}^{(l)} = \boldsymbol{g}\left([u_{i,k}^{(l)}, u_{j,k}^{(l)}]^T\right)$, $i, j = 1, \ldots, r$ and $\boldsymbol{p}_n^{(l)} \in \mathbb{R}^{n_p}$ are the regressor and the parameter vectors, respectively, and $\boldsymbol{g}(\cdot)$ is an arbitrary bivariate vector function, e.g., $\boldsymbol{g}(\boldsymbol{x}) = [x_1^2, x_2^2, x_1x_2, x_1, x_2, 1]^T$ that corresponding to the bivariate polynomial of the second degree.

An outline of the GMDH algorithm can be as follows [143, 185]:

*Step 1*: Determine all neurons (estimate their parameter vectors $\boldsymbol{p}_n^{(l)}$ with the training data set $\mathcal{T}$) whose inputs consist of all possible couples of input variables, i.e., $(r-1)r/2$ couples (neurons);

*Step 2*: Using a validation data set $\mathcal{V}$, not employed during the parameter estimation phase, select several neurons which are best fitted in terms of the chosen criterion;

*Step 3*: If the termination condition is fulfilled (either the network fits the data with desired a accuracy, or the introduction of new neurons did not induce a significant increase in the approximation abilities of the neural network), then STOP, otherwise use the outputs of the best-fitted neurons (selected in *Step 2*) to form the input vector for the next layer (see Fig. 7.13), and then go to *Step 1*.

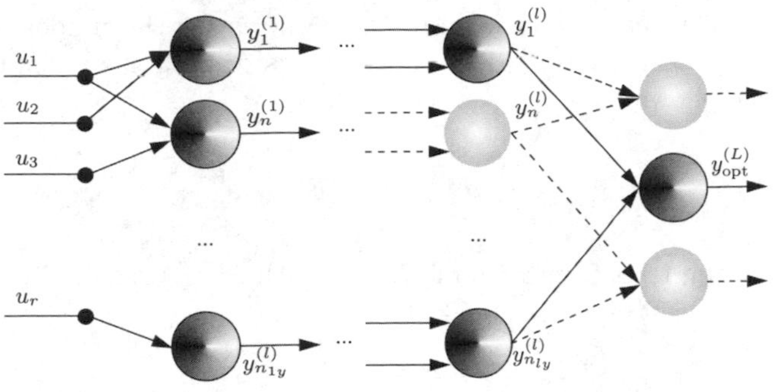


**Fig. 7.14.** Final structure of the GMDH neural network

To obtain the final structure of the network (Fig. 7.14), all unnecessary neurons are removed, leaving only those which are relevant to the computation of the model output. The procedure of removing the unnecessary neurons is the last stage of the synthesis of the GMDH neural network. The feature of the above algorithm is that the techniques for parameter estimation of linear-in-parameter models can be used during the realisation of *Step 1*. Indeed, since $\xi(\cdot)$ is invertible, the neuron (7.124) can relatively easily be transformed into a linear-in-parameter one.

### 7.2.1 Model Uncertainty in the GMDH Neural Network

The objective of system identification is to obtain a mathematical description of a real system based on input-output measurements. Irrespective of the identification method used, there is always the problem of model uncertainty, i.e., the model-reality mismatch. Even though the application of the GMDH approach to model structure selection can improve the quality of the model, the resulting structure is not the same as that of the system. It can be shown [119] that the application of the classic evaluation criteria like the Akaike Information Criterion (AIC) and the Final Prediction Error (FPE) [85, 121] can lead to the selection of inappropriate neurons and, consequently, to unnecessary structural errors.

Apart from the model structure selection stage, inaccuracy in parameter estimates also contributes to modelling uncertainty. Indeed, while applying the least-square method to parameter estimation of neurons (7.124), a set of restrictive assumptions has to be satisfied. The first, and the most controversial, assumption is that the structure of the neuron is the same as that of the system (no structural errors). In the case of the GMDH neural network, this condition is extremely difficult to satisfy. Indeed, neurons are created based on two input variables selected from $\boldsymbol{u}$ and hence it is impossible to eliminate the structural error. Another assumption concerns the transformation with $\xi^{-1}(\cdot)$. Let us consider the following system output signal:

$$y_k = \xi\left(\left(\boldsymbol{r}_{n,k}^{(l)}\right)^T \boldsymbol{p}_n^{(l)}\right) + v_{n,k}^{(l)}. \tag{7.125}$$

The use of linear-in-parameter estimation methods for the model (7.124), e.g., Least-Square Method (LSM) [161] requires transforming the output of the system (7.125) as follows:

$$\left(\boldsymbol{r}_{n,k}^{(l)}\right)^T \boldsymbol{p}_n^{(l)} = \xi^{-1}\left(y_k\right) - v_{n,k}^{(l)}. \tag{7.126}$$

Unfortunately, the transformation of (7.125) with $\xi^{-1}(\cdot)$ results in

$$\left(\boldsymbol{r}_{n,k}^{(l)}\right)^T \boldsymbol{p}_n^{(l)} = \xi^{-1}\left(y_k - v_{n,k}^{(l)}\right). \tag{7.127}$$

Thus, good results can only be expected when the noise $v_{n,k}^{(l)}$ magnitude is relatively small. The other assumptions are directly connected with the properties of the LSM, i.e., in order to attain an estimator $\hat{\boldsymbol{p}}_{n,k}^{(l)}$ of $\boldsymbol{p}_{n,k}^{(l)}$ for (7.124) which is unbiased and minimum variance [7] it is assumed that

$$\mathcal{E}\left[\boldsymbol{v}_n^{(l)}\right] = \mathbf{0}, \tag{7.128}$$

$$\mathrm{cov}\left[\boldsymbol{v}_n^{(l)}\right] = \left(\sigma_n^{(l)}\right)^2 \boldsymbol{I}. \tag{7.129}$$

The assumption (7.128) means that there are no structural errors (deterministic disturbances) and model uncertainty is described in a purely stochastic way (uncorrelated noise, cf. (7.129)). It must be pointed out that it is rather difficult to satisfy this condition in practice.

Let us suppose that, in some case, the conditions (7.128) and (7.129) are satisfied. Then it can be shown that $\hat{\boldsymbol{p}}_{n,k}^{(1)}$ (the parameter estimate vector for a neuron of the first layer) is unbiased and minimum variance [7]. Consequently, the neuron output in the first layer becomes the input to other neurons in the second layer. The system output estimate can be described by

$$\hat{\boldsymbol{y}}_n^{(l)} = \boldsymbol{R}_n^{(l)} \left[\left(\boldsymbol{R}_n^{(l)}\right)^T \boldsymbol{R}_n^{(l)}\right]^{-1} \left(\boldsymbol{R}_n^{(l)}\right)^T \boldsymbol{y}, \tag{7.130}$$

where $\boldsymbol{R}_n^{(l)} = [\boldsymbol{r}_{n,1}^{(l)}, \ldots, \boldsymbol{r}_{n,n_t}^{(l)}]^T$, $\boldsymbol{y} = [y_1, \ldots, y_{n_t}]^T$, and $\hat{\boldsymbol{y}}_n^{(l)} = [\hat{y}_{n,1}^{(l)}, \ldots, \hat{y}_{n,n_t}^{(l)}]^T$ represent the system output vector and its estimate. Apart from the situation in the first layer ($l = 1$), where the matrix $\boldsymbol{R}_n^{(l)}$ depends on $\boldsymbol{u}$, in the subsequent layers $\boldsymbol{R}_n^{(l+1)}$ depends on (7.130) and hence

$$\mathcal{E}\left[\left[\left(\boldsymbol{R}_n^{(l+1)}\right)^T \boldsymbol{R}_n^{(l+1)}\right]^{-1} \left(\boldsymbol{R}_n^{(l+1)}\right)^T \boldsymbol{v}_n^{(l+1)}\right] \neq \mathbf{0}. \tag{7.131}$$

That is why the parameter estimator in the next layers is biased and no minimum variance, i.e.,

$$\begin{aligned}\mathcal{E}\left[\hat{\boldsymbol{p}}_n^{(l+1)}\right] &= \mathcal{E}\left[\left[\left(\boldsymbol{R}_n^{(l+1)}\right)^T \boldsymbol{R}_n^{(l+1)}\right]^{-1}\left(\boldsymbol{R}_n^{(l+1)}\right)^T \boldsymbol{y}\right] \\ &= \mathcal{E}\left[\left[\left(\boldsymbol{R}_n^{(l+1)}\right)^T \boldsymbol{R}_n^{(l+1)}\right]^{-1}\left(\boldsymbol{R}_n^{(l+1)}\right)^T \left(\boldsymbol{R}_n^{(l+1)}\boldsymbol{p}_n^{(l+1)} + \boldsymbol{v}_n^{(l+1)}\right)\right] \\ &= \boldsymbol{p}_n^{(l+1)} + \mathcal{E}\left[\left[\left(\boldsymbol{R}_n^{(l+1)}\right)^T \boldsymbol{R}_n^{(l+1)}\right]^{-1}\left(\boldsymbol{R}_n^{(l+1)}\right)^T \boldsymbol{v}_n^{(l+1)}\right]. \end{aligned} \tag{7.132}$$

To settle this problem, the instrumental variable method or other methods listed in [170] can be employed. On the other hand, these methods provide only asymptotic convergence, and hence a large data set is usually required to obtain an unbiased parameter estimate.

### 7.2.2 Bounded-Error Approach

The problems detailed in the previous section clearly show that there is a need for the application of a parameter estimation method different than the LSM. Such a method should also be easily adaptable to the case of an uncertain regressor and it should overcome all of the remaining difficulties mentioned in Section 7.2.1. The subsequent part of this section gives an outline of such a method called the Bounded-Error Approach (BEA).

#### Bounded noise/disturbances

The usual statistical parameter estimation framework assumes that data are corrupted by errors which can be modelled as realisations of independent random variables, with a known or parameterised distribution. A more realistic approach is to assume that the errors lie between given prior bounds. This is the case, for example, for data collected with an analogue-to-digital converter or for measurements performed with a sensor of a given type. Such reasoning leads directly to the bounded-error approach [116, 170]. Let us consider the following system:

$$y_k = \left(\boldsymbol{r}_{n,k}^{(l)}\right)^T \boldsymbol{p}_n^{(l)} + v_{n,k}^{(l)}. \tag{7.133}$$

The problem is to obtain the parameter estimate $\hat{\boldsymbol{p}}_n^{(l)}$ as well as an associated parameter uncertainty required to design a robust fault detection system. In order to simplify the notation, the index ${}_n^{(l)}$ is omitted. The knowledge regarding the set of admissible parameter values allows obtaining the confidence interval of the model output which satisfies

$$y_k^N \leq y_k \leq y_k^M, \tag{7.134}$$

where $y_k^N$ and $y_k^M$ are respectively the minimum and maximum admissible values of the model output that are consistent with the input-output measurements of the system. Under the assumptions detailed in Section 7.2.1, the uncertainty of the neural network can be obtained according to [133].

In this work, it is assumed that $v_k$ is bounded as follows:

$$v_k^N \le v_k \le v_k^M, \tag{7.135}$$

while the bounds $v_k^N$ and $v_k^M$ ($v_k^N \ne v_k^M$) are known a priori. The idea underlying the bounded-error approach is to obtain a feasible parameter set [116]. This set can be defined as

$$\mathbb{P} = \left\{ \boldsymbol{p} \in \mathbb{R}^{n_p} \mid y_k - v_k^M \le \boldsymbol{r}_k^T \boldsymbol{p} \le y_k - v_k^N \ , k = 1, \ldots, n_t \right\}. \tag{7.136}$$

This set can be perceived as a region of the parameter space that is determined by $n_t$ pairs of hyperplanes:

$$\mathbb{P} = \bigcap_{k}^{n_t} \mathbb{S}_k, \tag{7.137}$$

where each pair defines the parameter strip

$$\mathbb{S}_k = \left\{ \boldsymbol{p} \in \mathbb{R}^{n_p} \mid y_k - v_k^M \le \boldsymbol{r}_k^T \boldsymbol{p} \le y_k - v_k^N \right\}. \tag{7.138}$$

Any parameter vector contained in $\mathbb{P}$ is a valid estimate of $\boldsymbol{p}$. In practice, the centre (in some geometrical sense) of $\mathbb{P}$ (cf. Fig. 7.15 for $n_p = 2$) is chosen as the parameter estimate $\hat{\boldsymbol{p}}$, e.g.,

$$\hat{p}_i = \frac{p_i^{\min} + p_i^{\max}}{2}, \quad i = 1, \ldots, n_p, \tag{7.139}$$

with

$$p_i^{\min} = \arg\min_{p \in \mathbb{P}} p_i, \quad i = 1, \ldots, n_p, \tag{7.140}$$

$$p_i^{\max} = \arg\max_{p \in \mathbb{P}} p_i, \quad i = 1, \ldots, n_p. \tag{7.141}$$

This is, of course, important when the task is to develop a neural network for which the knowledge regarding parameter uncertainty is not useful. In the presented approach, the nominal model is obtained in such a way that the knowledge regarding parameter uncertainty is used for fault detection purposes.

The problems (7.140) and (7.141) can be solved with the well-known linear programming techniques [116, 156], but when $n_t$ and/or $n_p$ are large, the computational cost may be significant. This constitutes the main drawback of the approach. One way out of this problem is to apply a technique where constraints are executed separately one after another [118], although this approach does not constitute a perfect remedy for the computational problem being considered. This means that the described BEA can be employed for tasks with a relatively small dimension, as is the case for GMDH neurons. In spite of the above-mentioned computational problems, the technique described in [118] was implemented and used in this work. The main difficulty associated with the BEA

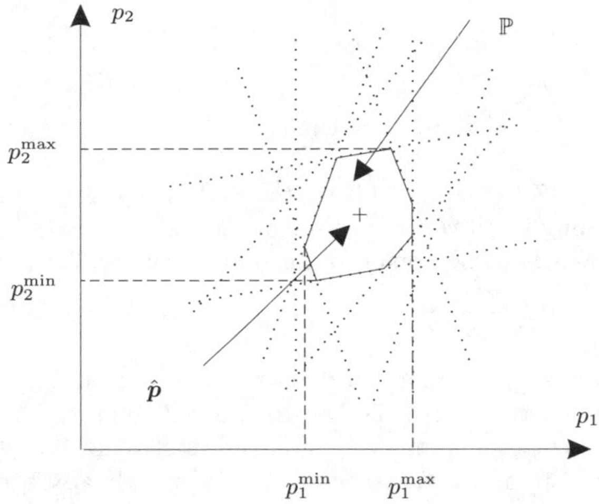


**Fig. 7.15.** Feasible parameter set

concerns a priori knowledge regarding the error bounds $v_k^N$ and $v_k^M$. However, these bounds can also be estimated [116] by assuming that $v_k^N = v^N$, $v_k^M = v^M$, and then suitably extending the unknown parameter vector $\boldsymbol{p}$ by $v^N$ and $v^M$. Determining the bounds can now be formulated as follows:

$$(v^N, v^M) = \arg \min_{v^M \geq 0,\ v^N \leq 0} v^M - v^N, \tag{7.142}$$

with respect to the following constraints:

$$y_k - v^M \leq \boldsymbol{r}_k^T \boldsymbol{p} \leq y_k - v^N \ , k = 1, \ldots, n_t. \tag{7.143}$$

In this section, the well-known simplex method was utilised to solve the problem (7.142). Then, knowing $v^N$ and $v^M$, the strategy described in [118] was employed.

## Model output uncertainty

The methodology described in Section 7.2.2 makes it possible to obtain the parameter estimate $\hat{\boldsymbol{p}}$ and the associated feasible parameter set $\mathbb{P}$. However, from a practical point of view, it is more convenient to obtain the system output confidence interval, i.e., the interval in which the "true" model output $y(k)$ can be found. This kind of knowledge makes it possible to obtain an adaptive threshold [57], and hence to develop a fault diagnosis scheme that is robust to model uncertainty.

Let $\mathbb{V}$ be the set of all vertices $\boldsymbol{p}^i$, $i = 1, \ldots, n_v$, describing the feasible parameter set $\mathbb{P}$ (cf. (7.137)). If there is no error in the regressor, then the problem of determining the model output confidence interval can be solved as follows:

$$y_{M,k}^N = \boldsymbol{r}_k^T \boldsymbol{p}_k^N \leq \boldsymbol{r}_k^T \boldsymbol{p} \leq \boldsymbol{r}_k^T \boldsymbol{p}_k^M = y_{M,k}^M, \tag{7.144}$$

where

$$\boldsymbol{p}_k^N = \arg\min_{\boldsymbol{p}\in\mathbb{V}} \boldsymbol{r}_k^T\boldsymbol{p}, \tag{7.145}$$

$$\boldsymbol{p}_k^M = \arg\max_{\boldsymbol{p}\in\mathbb{V}} \boldsymbol{r}_k^T\boldsymbol{p}. \tag{7.146}$$

The computation of (7.145) and (7.146) is realised by multiplying the parameter vectors corresponding to all vertices belonging to $\mathbb{V}$ by $\boldsymbol{r}_k^T$.

Since (7.144) describes a neuron output confidence interval, the system output will satisfy

$$\boldsymbol{r}_k^T\boldsymbol{p}_k^N + v_k^N \leq y_k \leq \boldsymbol{r}_k^T\boldsymbol{p}_k^M + v_k^M. \tag{7.147}$$

A more general case of (7.147) for neurons with a non-linear activation function will be considered in Section 7.2.3. The neuron output confidence interval defined by (7.144) and the corresponding system output confidence interval (7.147) are presented in Figs. 7.16 and 7.17, respectively. As has already been mentioned,

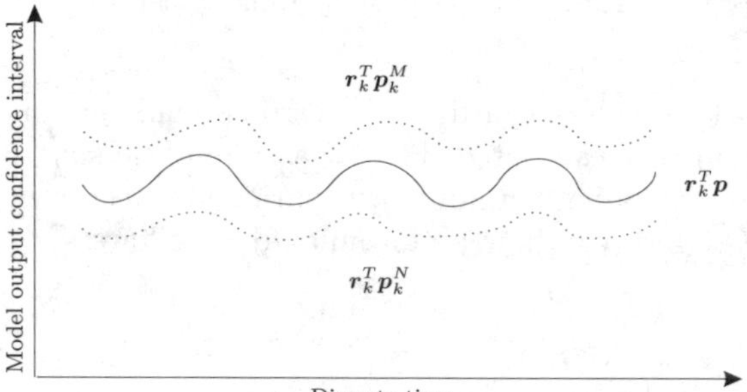


**Fig. 7.16.** Model output confidence interval for the error-free regressor

the neurons in the $l$th ($l > 1$) layer are fed with the outputs of the neurons from the $(l-1)$th layer. Since (7.144) describes the model output confidence interval, the parameters of the neurons in the layers have to be obtained with an approach that solves the problem of an uncertain regressor [116].

In order to modify the approach presented in Section 7.2.2, let us denote an unknown "true" value of the regressor $\boldsymbol{r}_{n,k}$ by a difference between a known (measured) value of the regressor $\boldsymbol{r}_k$ and the error in the regressor $\boldsymbol{e}_k$:

$$\boldsymbol{r}_{n,k} = \boldsymbol{r}_k - \boldsymbol{e}_k, \tag{7.148}$$

where it is assumed that the error $\boldsymbol{e}_k$ is bounded as follows:

$$e_{i,k}^N \leq e_{i,k} \leq e_{i,k}^M, \quad i = 1, \ldots, n_p. \tag{7.149}$$

Using (7.133) and substituting (7.148) into (7.149), one can define the region containing parameter estimates:

$$v_k^N - \boldsymbol{e}_k^T\boldsymbol{p} \leq y_k - \boldsymbol{r}_k^T\boldsymbol{p} \leq v_k^M - \boldsymbol{e}_k^T\boldsymbol{p}. \tag{7.150}$$

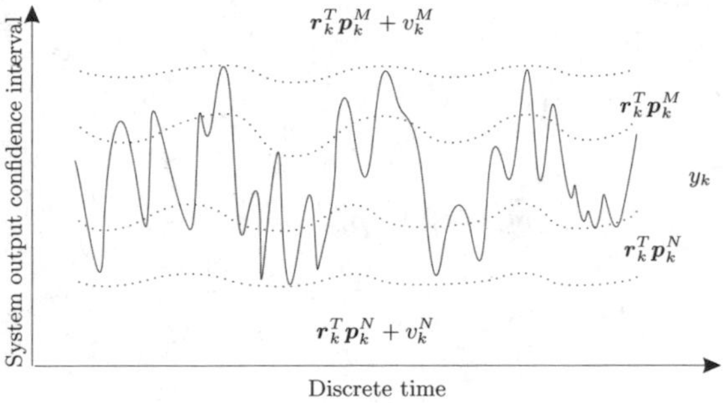


**Fig. 7.17.** System output confidence interval for the error-free regressor

Unfortunately, for the purpose of parameter estimation it is not enough to introduce (7.148) into (7.149). Indeed, the bounds of (7.150) depend also on the sign of each $p_i$. It is possible to directly obtain these signs only for models whose parameters have physical meaning [33]. For models such as GMDH neural networks this is rather impossible. In [116, Chapters 17 and 18], the authors proposed some heuristic techniques, but these drastically complicate the problem (7.150) and do not seem to guarantee that these signs will be obtained properly. Bearing in mind the fact that the neuron (7.124) contains only a few parameters, it is possible to replace them by

$$p_i = p_i' - p_i'', \quad p_i', p_i'' \geq 0, \quad i = 1, \ldots, n_p. \tag{7.151}$$

Although the above solution is very simple, it doubles the number of parameters, i.e., instead of estimating $n_p$ parameters it is necessary to do so for $2n_p$ parameters. In spite of that, this technique is very popular and widely used in the literature [52, 116]. Due to the above solution, (7.150) can be modified as follows:

$$\begin{aligned} &v_k^N - \left(e_k^M\right)^T p' + \left(e_k^N\right)^T p'' \\ &\leq y_k - r_k^T (p' - p'') \leq \\ &v_k^M - \left(e_k^N\right)^T p' + \left(e_k^M\right)^T p''. \end{aligned} \tag{7.152}$$

This transformation makes it possible to employ, with a minor modification, the approach described in Section 7.2.2. The difference is that the algorithm processes each constraint (associated with a pair of hyperplanes defined with (7.152)) separately. The reason for such a modification is that the hyperplanes are not parallel [34].

The proposed modification of the BEA makes it possible to estimate the parameter vectors of the neurons from the $l$th, $l > 1$, layer. In the case of an error in the regressor, using (7.152) it can be shown that the model output confidence interval has the following form:

$$y_{M,k}^N(p_k'^N, p_k''^N) \leq r_n^T p \leq y_{M,k}^M(p_k'^M, p_k''^M), \tag{7.153}$$

where

$$y_{M,k}^{N}(\boldsymbol{p}_k'^{N}, \boldsymbol{p}_k''^{N}) = \left(\boldsymbol{r}_k - \boldsymbol{e}_k^{M}\right)^T \boldsymbol{p}_k'^{N} + \left(\boldsymbol{e}_k^{N} - \boldsymbol{r}_k\right)^T \boldsymbol{p}_k''^{N}, \tag{7.154}$$

$$y_{M,k}^{M}(\boldsymbol{p}_k'^{M}, \boldsymbol{p}_k''^{M}) = \left(\boldsymbol{r}_k - \boldsymbol{e}_k^{N}\right)^T \boldsymbol{p}_k'^{M} + \left(\boldsymbol{e}_k^{M} - \boldsymbol{r}_k\right)^T \boldsymbol{p}_k''^{M}, \tag{7.155}$$

and

$$\left(\boldsymbol{p}_k'^{N}, \boldsymbol{p}_k''^{N}\right) = \arg \min_{(\boldsymbol{p}_k', \boldsymbol{p}_k'') \in \mathbb{V}} y_{M,k}^{N}(\boldsymbol{p}_k', \boldsymbol{p}_k''), \tag{7.156}$$

$$\left(\boldsymbol{p}_k'^{M}, \boldsymbol{p}_k''^{M}\right) = \arg \max_{(\boldsymbol{p}_k', \boldsymbol{p}_k'') \in \mathbb{V}} y_{M,k}^{M}(\boldsymbol{p}_k', \boldsymbol{p}_k''). \tag{7.157}$$

Using (7.153), it is possible to obtain the system output confidence interval:

$$y_{M,k}^{N}(\boldsymbol{p}_k'^{N}, \boldsymbol{p}_k''^{N}) + v_k^{N} \le y_k \le y_{M,k}^{M}(\boldsymbol{p}_k'^{M}, \boldsymbol{p}_k''^{M}) + v_k^{M}. \tag{7.158}$$

### 7.2.3 Synthesis of the GMDH Neural Network Via the BEA

In order to adapt the approach of Section 7.2.2 to parameter estimation of (7.124), it is necessary to transform the relation

$$v_k^{N} \le y_k - \xi\left(\left(\boldsymbol{r}_{n,k}^{(l)}\right)^T \boldsymbol{p}_n^{(l)}\right) \le v_k^{M} \tag{7.159}$$

in such a way as to avoid the problems detailed in Section 7.2.1. In this case, it is necessary to assume that

1. $\xi(\cdot)$ is continuous and bounded, i.e.,

$$\forall\, x \in \mathbb{R} \;: a < \xi(x) < b; \tag{7.160}$$

2. $\xi(\cdot)$ is monotonically increasing, i.e.,

$$\forall\, x, y \in \mathbb{R} \;: x \le y \text{ iff } \xi(x) \le \xi(y). \tag{7.161}$$

Now it is easy to show that

$$y_k - v_k^{M} \le \xi\left(\left(\boldsymbol{r}_{n,k}^{(l)}\right)^T \boldsymbol{p}_n^{(l)}\right) \le y_k - v_k^{N}, \tag{7.162}$$

and then

$$\xi^{-1}\left(y_k - v_k^{M}\right) \le \left(\boldsymbol{r}_{n,k}^{(l)}\right)^T \boldsymbol{p}_n^{(l)} \le \xi^{-1}\left(y_k - v_k^{N}\right). \tag{7.163}$$

As was pointed out in Section 7.2.2, an error in the regressor must be taken into account during the design procedure of the neurons from the second and subsequent layers. Indeed, by using (7.144) in the first layer and (7.153) in the subsequent ones it is possible to obtain the bounds of the output (7.124) and the bounds of the regressor error (7.135), whilst the known value of the regressor

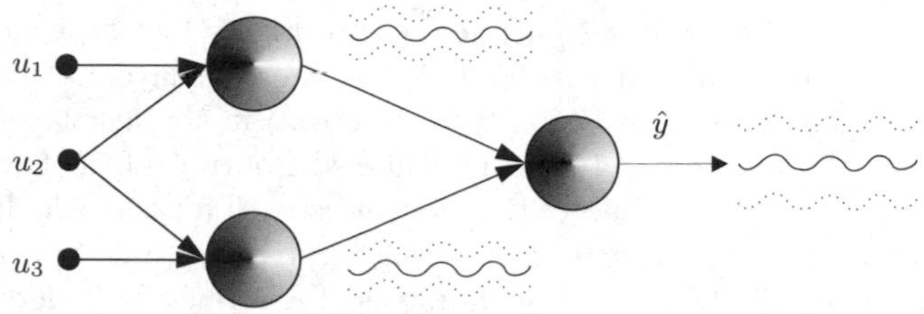


**Fig. 7.18.** Propagation of model uncertainty (dotted lines), model response (continuous line)

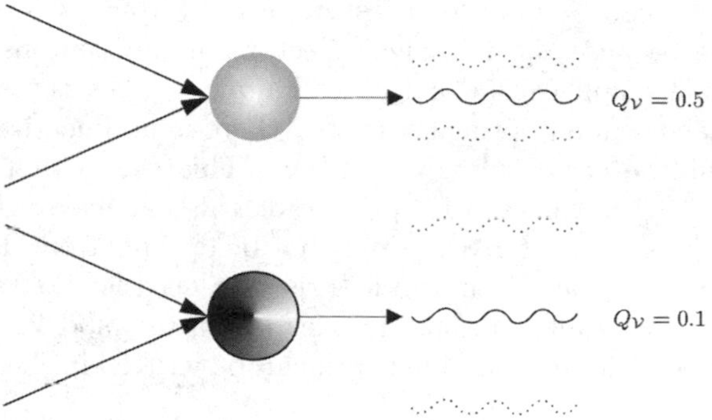


**Fig. 7.19.** Problem of an incorrect selection of a neuron

should be computed by using the parameter estimates $\hat{\boldsymbol{p}}_n^{(l)}$. Note that the processing errors of the neurons, which are described by the model output confidence interval (7.153), can be propagated and accumulated during the introduction of new layers (Fig. 7.18). This unfavourable phenomenon can be reduced by the application of an appropriate selection method [136]. Selection methods in GMDH neural networks play the role of a mechanism of structural optimisation at the stage of constructing a new layer of neurons. Only well-performing neurons are preserved to build a new layer. During the selection, neurons which have too large a value of the defined evaluation criteria [85, 121, 136] are rejected based on chosen selection methods. Unfortunately, as was mentioned in Section 7.2.1, the application of the classic evaluation criteria like the Akaike Information Criterion (AIC) and the Final Prediction Error (FPE) [85, 121] during network synthesis may lead to the selection of an inappropriate structure of the GMDH neural network. This follows from the fact that the above criteria do not take into account modelling uncertainty. In this way, neurons with small values of the classic quality indexes $Q_{\mathcal{V}}$ but with large uncertainty (Fig. 7.19) can be obtained. In order to overcome this difficulty, a new evaluation criterion of the neurons has been introduced in [185], i.e.,

$$Q_{\mathcal{V}} = \frac{1}{n_{\mathcal{V}}} \sum_{k=1}^{n_v} \left| \left(y_{M,k}^M + v_k^M\right) - \left(y_{M,k}^N + v_k^N\right) \right|, \tag{7.164}$$

where $n_{\mathcal{V}}$ is the number of input-output measurements for the validation data set, $y_k^M$ and $y_k^N$ are calculated with (7.144) for the first layer or with (7.154)–(7.155) for the subsequent ones. Finally, the neuron in the last layer that gives the smallest processing error (7.164) constitutes the output of the GMDH neural network, while model uncertainty of this neuron is used for the calculation of the system output confidence interval. It is therefore possible to design the so-called adaptive threshold [57], which can be employed for robust fault detection.

### Further improvement of model quality

One of the main advantages of GMDH neural networks is the fact that the BEA for linear systems can be applied to estimate the parameters of each neuron. This is possible because the parameter vectors of the neurons are estimated independently. The application of this technique implies that the parameter vectors are obtained in an optimal way, i.e., there is no linearisation. However, optimality should be perceived as a local one. This means that the parameter vector associated with a neuron is optimal for this particular neuron only. On the other hand, this parameter vector may not be optimal from the point of view of the entire network. Such circumstances rise the need for the retraining of the GMDH neural network after automatic selection of the model structure.

Assume that the GMDH neural network can be written in the following form:

$$y_M = g\left(\boldsymbol{p}_1^{(1)}, \ldots \boldsymbol{p}_{n_{1y}}^{(1)}, \ldots \boldsymbol{p}_1^{(L)}, \ldots \boldsymbol{p}_{n_{Ly}}^{(L)}, \boldsymbol{u}\right), \tag{7.165}$$

and $g(\cdot)$ stands for the neural network structure obtained with the GMDH approach, $L$ is the number of layers, $n_{iy}$ is the number of neurons in the $i$th layer. After the procedure of designing the GMDH neural network, described in Section 7.2.3, the following sequence of estimates is obtained: $\hat{\boldsymbol{p}}_1^{(1)}, \ldots, \hat{\boldsymbol{p}}_{n_y}^{(1)}, \ldots, \hat{\boldsymbol{p}}_1^{(L)}, \ldots, \hat{\boldsymbol{p}}_{n_y}^{(L)}$. With each of these estimates, a feasible parameter set is associated, i.e.,

$$\boldsymbol{A}_{i,j}\boldsymbol{p}_i^{(j)} \leq \boldsymbol{b}_{i,j}, \quad i = 1, \ldots, n_{jy}, \; j = 1, \ldots, L, \tag{7.166}$$

where the matrix $\boldsymbol{A}_{i,j}$ and the vector $\boldsymbol{b}_{i,j}$ are used for describing the feasible parameter sets of each neuron of (7.165). These feasible parameter sets are known after automatic selection of the model structure and the estimation of its parameters. This makes it possible to formulate the parameter estimation task in a global sense. Indeed, it can be defined as a constrained optimisation problem of

$$Q_T = \frac{1}{n_t} \sum_{k=1}^{n_t} |y_k - \hat{y}_k|, \tag{7.167}$$

while

$$\hat{y}_k = g\left(\hat{\boldsymbol{p}}_1^{(1)}, \ldots \hat{\boldsymbol{p}}_{n_y}^{(1)}, \ldots \hat{\boldsymbol{p}}_1^{(L)}, \ldots \hat{\boldsymbol{p}}_{n_y}^{(L)}, \boldsymbol{u}_k\right), \tag{7.168}$$

while the constraints are given by (7.166). The choice of the $l_1$-norm in (7.167) is motivated by the fact that the properties of the $l_1$ estimator are very similar to these of the bounded error approach [170]. However, it should be pointed out that other criteria may successfully be implemented as well.

The solution of (7.167) can be obtained with optimisation techniques related to non-linear $l_1$ optimisation [170] as well as with specialised evolutionary algorithms [11]. The choice of the optimisation routine should be motivated by the fact that the global optimisation problem (7.167) has to be solved. This means that the application of the classic methods boils down to the trial-and-error procedure, which is ineffective. Based on numerous computer experiments, it has been found that the extremely simple ARS algorithm [170] is especially well suited for that purpose. Apart from its simplicity, the algorithm decreases the chance to get stuck in a local optimum and hence it may give a global minimum of (7.167).

### 7.2.4 Robust Fault Detection with the GMDH Model

The purpose of this section is to show how to develop an adaptive threshold with the GMDH model and some knowledge regarding its uncertainty. Since the residual is

$$z_k = y_k - \hat{y}_k, \tag{7.169}$$

then, as a result of substituting (7.169) into (7.158), the adaptive threshold can be written as

$$y^N_{M,k}(\boldsymbol{p}'^N_k, \boldsymbol{p}''^N_k) - y_k + v^N_k \leq z_k \leq y^M_{M,k}(\boldsymbol{p}'^M_k, \boldsymbol{p}''^M_k) - y_k + v^M_k. \tag{7.170}$$

The principle of fault detection with the developed adaptive threshold is shown in Fig. 7.20.

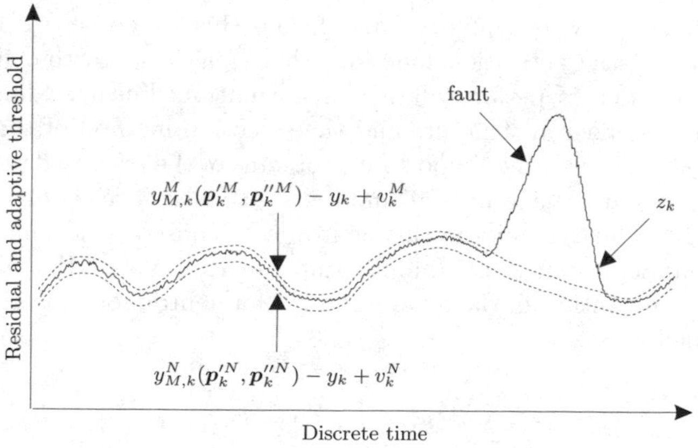


**Fig. 7.20.** Fault detection with an adaptive threshold developed with the proposed approach

### 7.2.5 Industrial Application

The purpose of the present section is to show the effectiveness of the proposed approach in the context of system identification and fault detection with the DAMADICS benchmark introduced in Section 6.4.2. Based on the actuator benchmark definition [12, 36], two GMDH models were designed. These models describe the behaviour of the valve actuator and can be labelled as the juice flow model $F = r_F(X, P_1, P_2, T_1)$ and the servomotor rod displacement model $X = r_X(C_V, P_1, P_2, T_1)$, where $r_F$ and $r_X$ stand for the modelled relation between the inputs $F, X$ and the outputs $C_V, X, P_1, P_2, T_1$.

The real data used for system identification and the fault detection procedure were collected on 17th November 2001. A detailed description regarding the data and the artificially introduced faults can be found in Tab. 7.2.

**Table 7.2.** List of data sets

| Fault | Range (samples) | Fault/data description |
|---|---|---|
| No fault | 1–10000 | Training data set |
| No fault | 10001–20000 | Validation data set |
| $f_{16}$ | 57475–57530 | Positioner supply pressure drop |
| $f_{17}$ | 53780–53794 | Unexpected pressure drop across valve |
| $f_{18}$ | 54600–54700 | Fully or partly opened bypass valves |
| $f_{19}$ | 55977–56015 | Flow rate sensor fault |

Unfortunately, the data turned out to be sampled too fast. Thus, every 10th value was picked, resulting in the $n_t = 1000$ training and $n_\mathcal{V} = 1000$ validation data sets. Moreover, the output data should be transformed taking into account the response range of the neuron output. In this section, hyperbolic tangent activation functions were employed and hence this range is $[-1, 1]$. To avoid the saturation of the activation function, this range was further decreased to $[-0.8, 0.8]$. In order to perform data transformation, linear scaling was used. The choice of the neuron structure and the selection method of the neurons in the GMDH network are other important problems of the proposed technique. For that purpose, dynamic neurons [120] and the so-called soft selection method [136] were employed. The dynamics in this neuron are realised by the introduction of a linear dynamic system – an Infinite Impulse Response (IIR) filter. As has previously been mentioned, the quality index of a neuron for the validation data set was defined as

$$Q_\mathcal{V} = \frac{1}{n_v} \sum_{k=1}^{n_v} \left| \left(y_{M,k}^M + v_k^M\right) - \left(y_{M,k}^N + v_k^N\right) \right|, \tag{7.171}$$

where $y_{M,k}^M$ and $y_{M,k}^N$ are calculated with (7.144) for the first layer or with (7.154)–(7.155) for the subsequent ones. Table 7.3 presents the evolution of (7.171) for the subsequent layers, i.e., these values are obtained for the best

**Table 7.3.** Evolution of $Q_{\mathcal{V}}$ and $B_{\mathcal{V}}$ for the subsequent layers

| Layer | $r_F(\cdot)$ $Q_{\mathcal{V}}$ | $r_F(\cdot)$ $B_{\mathcal{V}}$ | $r_X(\cdot)$ $Q_{\mathcal{V}}$ | $r_X(\cdot)$ $B_{\mathcal{V}}$ |
|---|---|---|---|---|
| 1 | 1.5549 | 0.3925 | 0.5198 | 0.0768 |
| 2 | 1.5277 | 0.3681 | 0.4914 | 0.0757 |
| 3 | 1.5047 | 0.3514 | 0.4904 | 0.0762 |
| 4 | 1.4544 | 0.3334 | 0.4898 | 0.0750 |
| 5 | 1.4599 | 0.3587 | 0.4909 | 0.0748 |

performing neurons in a particular layer. Additionally, for the sake of comparison, the results based on the classic quality index [121],

$$B_{\mathcal{V}} = \frac{1}{n_v}\sum_{k=1}^{n_v}\left|y_k - \hat{y}_k\right|, \tag{7.172}$$

are presented as well.

The results presented in Tab. 7.3 clearly show that the gradual decrease $Q_{\mathcal{V}}$ occurs when a new layer is introduced. This follows from the fact that the introduction of a new neuron increases the complexity of the model as well as its modelling abilities. On the other hand, if the model is too complex, then the quality index $Q_{\mathcal{V}}$ increases. This situation occurs, for both $F = r_F(\cdot)$ and $X = r_X(\cdot)$, when the 5th layer is introduced. This means that GMDH neural networks corresponding to $F = r_F(\cdot)$ and $X = r_X(\cdot)$ should have 4 layers. From Tab. 7.3, it can be also seen that the application of the quality index $B_{\mathcal{V}}$ gives similar results for $F = r_F(\cdot)$, i.e., the same number of layers was selected, whilst for $X = r_X(\cdot)$ it leads to the selection of too simple a structure, i.e., a neural network with only two layers is selected. This implies that the quality index $Q_{\mathcal{V}}$ makes it possible to obtain a model with a smaller uncertainty. In order to achieve the final structure of $F = r_F(\cdot)$ and $X = r_X(\cdot)$, all unnecessary neurons were removed, leaving only those that are relevant for the computation of the model output. The final structures of GMDH neural networks are presented in Figs. 7.21 and 7.22. From Fig. 7.22, it can be seen that the input variable $P_2$, was removed during the model development procedure. Nevertheless, the quality index $Q_{\mathcal{V}}$ achieved a relatively low level. It can be concluded that $P_2$ has a

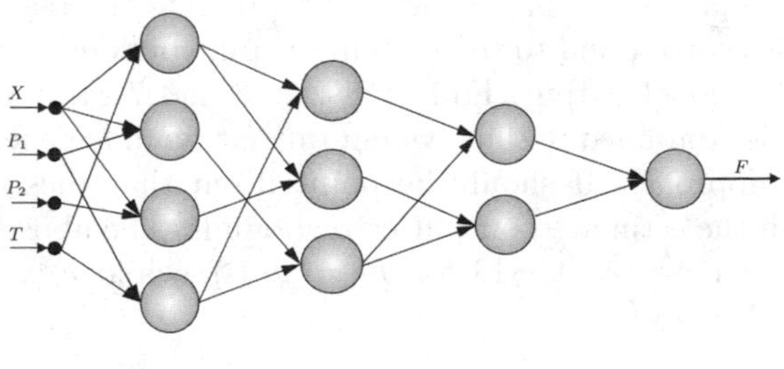


**Fig. 7.21.** Final structure of $F = r_F(\cdot)$

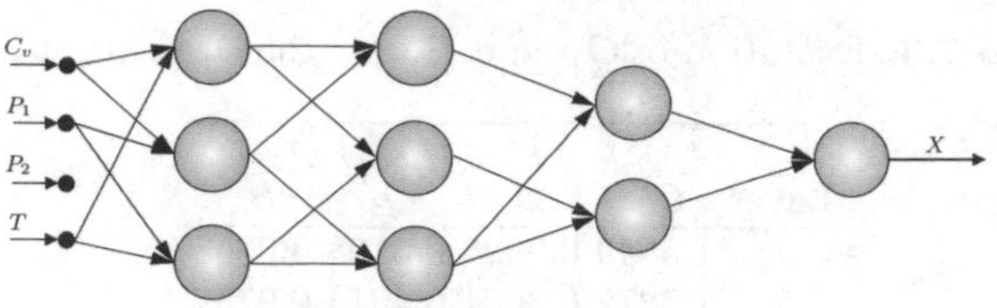


**Fig. 7.22.** Final structure of $X = r_X(\cdot)$

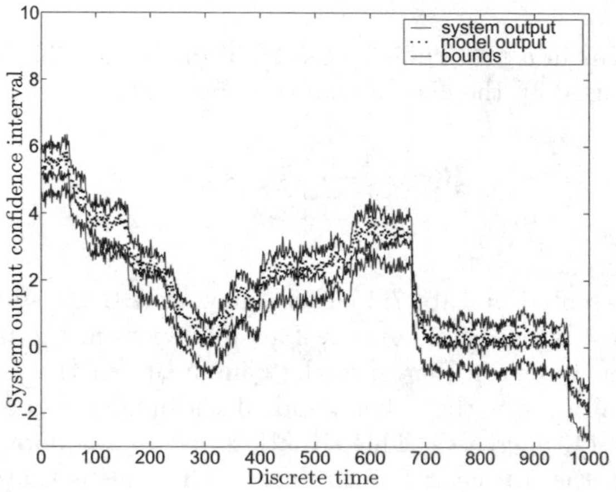


**Fig. 7.23.** Model and system output as well as the corresponding system output confidence interval for $F = r_F(\cdot)$

relatively small influence on the servomotor rod displacement $X$. This is an example of structural errors that may occur during the selection of neurons in the layer of the GMDH network. On the other hand, the proposed fault detection scheme is robust to such errors. This is because they are taken into account during the calculation of a model output confidence interval.

Figs. 7.23 and 7.24 present the modelling abilities of the obtained models $F = r_F(\cdot)$ and $X = r_X(\cdot)$ as well as the corresponding system output confidence interval obtained with the proposed approach for the validation data set.

For the reader's convenience, Fig. 7.25 presents a selected part of Fig. 7.23 for $k = 400 - 500$ samples. The thick solid line represents the real system output, the thin solid lines correspond to the system output confidence interval, and the dashed line is the model output. From Figs. 7.23 and 7.24, it is clear that the system response is contained within system output bounds generated according to the proposed approach. It should be pointed out that these system bounds are designed with the estimated output error bounds. The above estimates were $v_{n_t}^N = -0.8631$ and $v_{n_t}^M = 0.5843$ for $F = r_F(\cdot)$, while $v_{n_t}^N = -0.2523$ and $v_{n_t}^M = 0.2331$ for $X = r_X(\cdot)$.

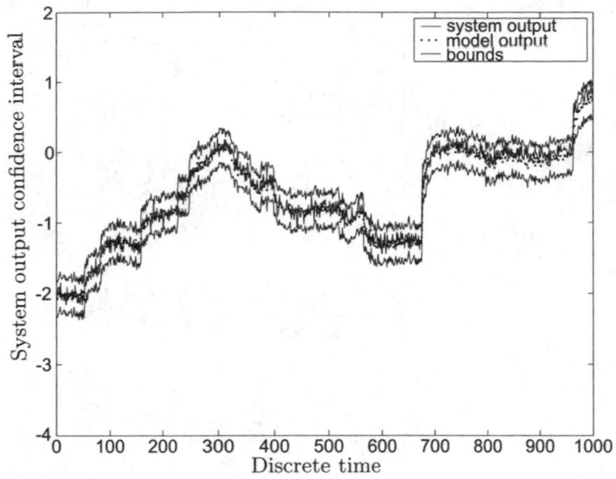


**Fig. 7.24.** Model and system output as well as the corresponding system output confidence interval for $X = r_X(\cdot)$

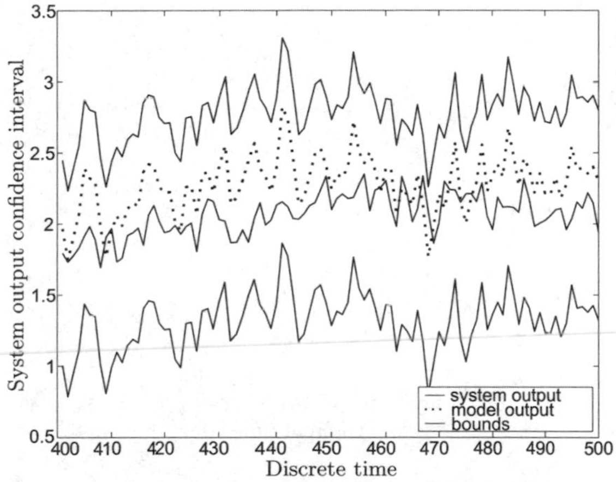


**Fig. 7.25.** Selected part of Fig. 7.23

As has already been mentioned, the quality of the GMDH model can be further improved with the application of the technique described in Section 7.2.3. This technique can be perceived as the retraining method for the network. For the valve actuator being considered, it was profitable to utilise the retraining technique for the model $F = r_F(\cdot)$. As a result, the quality index (7.172) was decreased from 0.3334 to 0.2160 (cf. Tab. 7.3). These results as well as the comparison of Figs. 7.25 and 7.26 justify the need for the retraining technique proposed in Section 7.2.3.

The main objective of this application study was to develop a fault detection scheme for the valve actuator being considered. Since both $F = r_F(\cdot)$ and $X = r_X(\cdot)$ were designed with the approach proposed in Section 7.2.3, it is possible

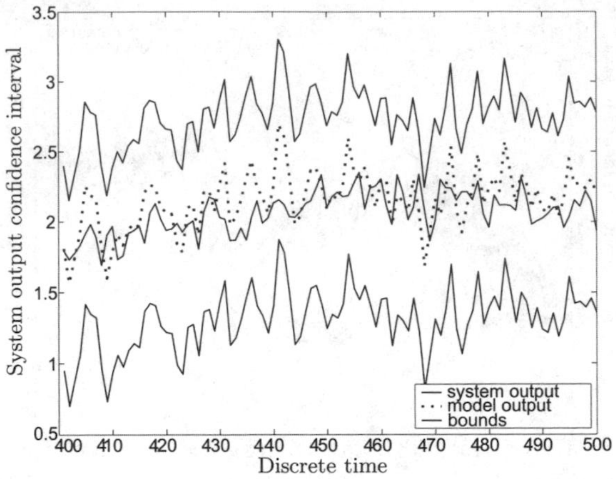


**Fig. 7.26.** Response of $F = r_F(\cdot)$ after retraining

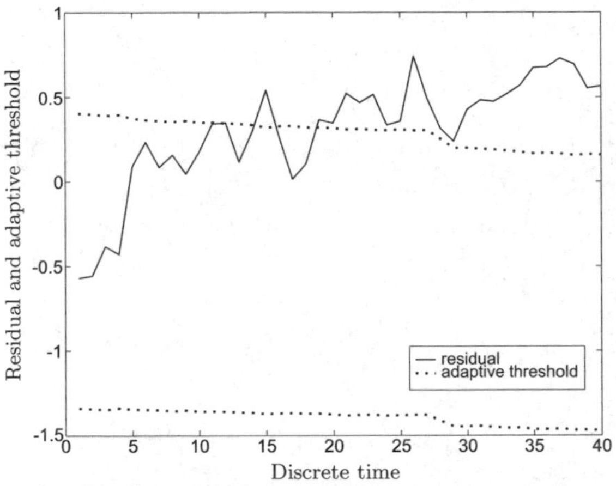


**Fig. 7.27.** Residual for the fault $f_{16}$

to employ them for robust fault detection. This task can be realised according to the rules described in Section 7.2.4. Figs. 7.27-7.30 present the residuals and their bounds for the faulty data.

From these results it can be seen that it is possible to detect all four faults, although the fault $f_{18}$ was detected 18 s after its occurrence. This is caused by the relative insensitivity of the obtained model to this particular fault. The results presented so far were obtained with data from a real system. It should also be pointed out that, within the framework of the actuator benchmark [12], data for only four general faults $f_{16} - f_{19}$ were available.

In order to provide a more comprehensive and detailed application study of the proposed fault diagnosis scheme, a MATLAB SIMULINK actuator model was

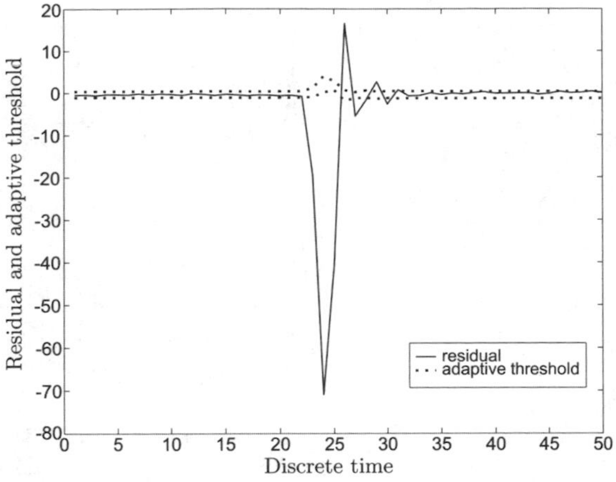


**Fig. 7.28.** Residual for the fault $f_{17}$

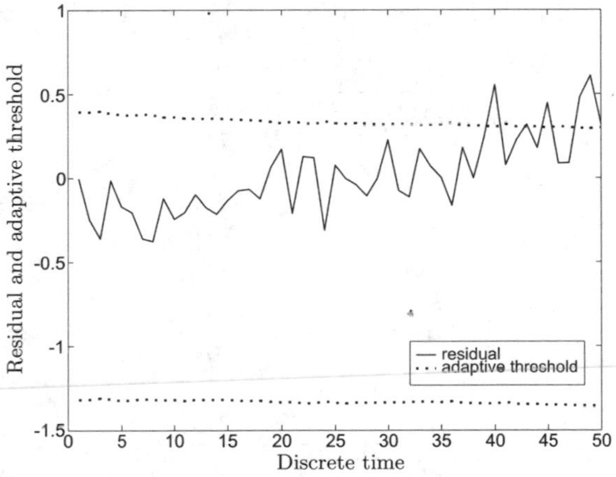


**Fig. 7.29.** Residual for the fault $f_{18}$

employed. This tool makes it possible to generate data for 19 different faults. Table 7.4 shows the results of fault detection. It should be pointed out that both abrupt and incipient faults were considered. As can be seen, the abrupt faults presented in Tab. 7.4 can be regarded as small, medium and big according to the benchmark description [12]. The notation given in Tab. 7.4 can be explained as follows: $ND$ means that it is impossible to detect a given fault, $D$ means that it is possible to detect a fault. From the results presented in Tab. 7.4, it can be seen that it is impossible to detect the faults $f_5$, $f_9$, and $f_{14}$. Moreover, some small and medium faults cannot be detected, i.e., $f_8$ and $f_{12}$. This situation can be explained by the fact that the effect of these faults is at the same level as the effect of noise.

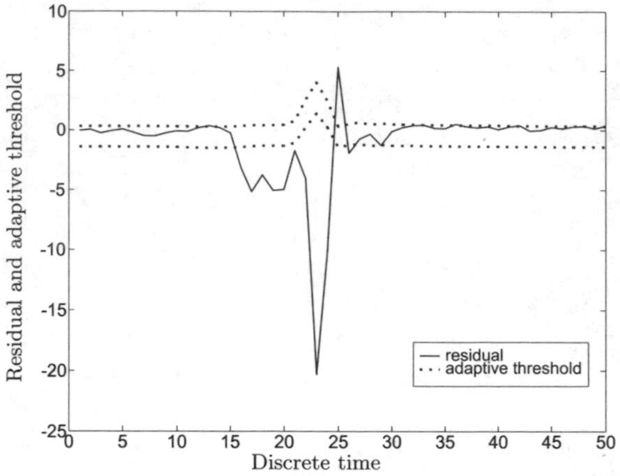


**Fig. 7.30.** Residual for the fault $f_{19}$

**Table 7.4.** Results of fault detection (S – small, M – medium, B – big, I – incipient)

| ***F*** | **S** | **M** | **B** | **I** |
|---|---|---|---|---|
| $f_1$ | $D$ | $D$ | $D$ | |
| $f_2$ | | | $D$ | $D$ |
| $f_3$ | | | | $D$ |
| $f_4$ | | | | $D$ |
| $f_5$ | | | | $ND$ |
| $f_6$ | | | | $D$ |
| $f_7$ | $D$ | $D$ | $D$ | |
| $f_8$ | $ND$ | $ND$ | $D$ | |
| $f_9$ | | | | $ND$ |
| $f_{10}$ | $D$ | $D$ | $D$ | |
| $f_{11}$ | | | $D$ | $D$ |
| $f_{12}$ | $ND$ | $ND$ | $D_X$ | |
| $f_{13}$ | $D$ | $D$ | $D$ | $D$ |
| $f_{14}$ | $ND$ | $ND$ | $ND$ | |
| $f_{15}$ | | | $D$ | |
| $f_{16}$ | $D$ | $D$ | $D$ | |
| $f_{17}$ | | | $D$ | $D$ |
| $f_{18}$ | $D$ | $D$ | $D$ | $D$ |
| $f_{19}$ | $D$ | $D$ | $D$ | |

## 7.3 Concluding Remarks

The present chapter presents two complete design procedures concerning the application of neural networks to robust fault detection. Section 7.1 shows how to settle such a challenging task with a multi-layer perceptron. In particular, it was shown how to describe model uncertainty of the MLP with statistical techniques. Subsequently, two algorithms that can be used for decreasing such a model uncertainty with the use of experimental design theory were presented and described in detail. It was also shown how to use the resulting knowledge about model uncertainty for robust fault detection with the so-called adaptive threshold. This section presents numerical examples that show all profits that can be gained while using the proposed algorithms. An industrial application case study concerning the DAMADICS benchmark was also presented in this section. In particular, it was shown how to design an experiment for the MLP being the model of the valve actuator. It was also shown how to use the resulting model and the knowledge about its uncertainty for a robust fault detection study based on a set of select faults.

A similar task was realised in Section 7.2. Starting from a set of input-output measurements of the system, it was shown how to estimate the parameters and the corresponding uncertainty of a neuron via the BEA. The methodology developed for parameter and uncertainty estimation of a neuron makes it possible to formulate an algorithm that allows obtaining a neural network with a relatively small modelling uncertainty. Subsequently, a complete design procedure of a neural network was proposed and carefully described. All the hard computation regarding the design of the GMDH neural network are performed offline and hence the problem regarding the time–consuming calculations is not of paramount importance. Based on the GMDH neural network, a novel robust fault detection scheme was proposed which supports diagnostic decisions. Similarly as in Section 7.1, the presented approach was tested with the DAMADICS benchmark.

The experimental results presented in this chapter clearly show all profits that can be gained while using the proposed neural network-based fault detection schemes. It is worth noting that they can be successfully employed instead of the classic techniques, e.g., the unknown input observers described in Chapter 4. Indeed, the robustness of the presented fault detection tools makes them useful for solving challenging design problems that arise in engineering practice.

# 8. Conclusions and Future Research Directions

From the point of view of engineering, it is clear that providing fast and reliable fault detection and isolation is an integral part of control design, particularly as far as the control of complex industrial systems is considered.

Unfortunately, most of such systems exhibit non-linear behaviour, which makes it impossible to use the well-developed techniques for linear systems. If it is assumed that the system is linear, which is not true in general, and even if robust techniques for linear systems are used (e.g., unknown input observers), it is clear that such an approximation may lead to unreliable fault detection and, consequently, early indication of faults which are developing is rather impossible. Such a situation increases the probability of the occurrence of faults, which can be extremely serious in terms of economic losses, environmental impact, or even human mortality. Indeed, robust techniques are able to tolerate a certain degree of model uncertainty. In other words, they are not robust to everything, i.e., are robust to an arbitrary degree of model uncertainty. This real world development pressure creates the need for new techniques which are able to tackle fault diagnosis of non-linear systems. In spite of the fact that the problem has been attacked from various angles by many authors and a number of relevant results have already been reported in the literature, there is no general framework which can be simply and conveniently applied to maintain fault diagnosis for non-linear systems.

As was indicated in Part I, there are two general fault diagnosis frameworks, which divide the existing approaches into two distinct categories, i.e.,

- Analytical techniques (Chapter 2);
- Soft computing techniques (Chapter 3).

Moreover, within the first category, the three main approaches can be distinguished:

- Parameter estimation;
- Parity relation;
- Observers.

M. Witczak: Model. and Estim. Strat. for Fault Diagn. of Non-Linear Syst. LNCIS 354, pp. 185–190, 2007.
springerlink.com 

As was underlined in Chapter 2, observers are immensely popular as residual generators for fault detection (and, consequently, for fault isolation) of both linear and non-linear dynamic systems. Their popularity lies in the fact that they can also be employed for control purposes. There are, of course, many different observers which can be applied to non-linear, and especially non-linear deterministic systems, and the best known of them were briefly reviewed in Section 2.2.3. Logically, the number of "real world" applications (not only simulated examples) should proliferate, yet this is not the case. It seems that there are two main reasons why strong formal methods are not accepted in engineering practice. First, the design complexity of most observers for non-linear systems does not encourage engineers to apply them in practice. Second, the application of observers is limited by the need for non-linear state-space models of the system being considered, which is usually a serious problem in complex industrial systems. This explains why most of the examples considered in the literature are devoted to simulated or laboratory systems, e.g., the celebrated three- (two- or even four-) tank system, an inverted pendulum, a travelling crane, etc. The above discussion clearly justifies the need for simpler observer structures, which can be obtained by solving the following problems (see Section 2.4):

*Problem 1:* Improvement of the convergence of linearisation-based observers;
*Problem 2:* Simplification of linearisation-free observers.

These two problems have to be solved under an additional robustness condition, i.e., the observers being designed have to ensure robustness to model uncertainty.

The remaining task concerns non-linear state-space model design for observer-based fault diagnosis and can be formulated as follows:

*Problem 3:* Development of a design technique for non-linear state-space models.

As was mentioned in Chapter 3, challenging design problems arise regularly in modern fault diagnosis systems. Unfortunately, the classic analytical techniques often cannot provide acceptable solutions to such difficult tasks. If this is the case, one possible approach is to use soft computing-based fault diagnosis approaches, which can be divided into three categories:

- Neural networks;
- Fuzzy logic-based techniques;
- Evolutionary algorithms.

Apart from the unquestionable appeal of soft computing approaches, there is a number of design issues that can be described by (see Section 3.3):

*Problem 4*: Integration of analytical and soft computing FDI techniques;
*Problem 5*: Development of robust soft computing-based FDI techniques.

Finally, the last problem was exposed in Chapter 5 and can be formulated as follows:

*Problem 6*: Need for the development of active fault diagnosis for non-linear systems.

Although partial solutions to *Problems 1-6* are scattered over many papers and a number of book chapters, there is no work that summarises all of these results in a unified framework.

Thus, one original objective of this book was to present selected modelling and estimation strategies for solving the challenging *Problems 1-6* in a unified framework.

Other objectives, perceived as solutions to *Problems 1-6*, are presented in the form of a concise summary of the contributions provided by this book to the state-of-the-art of modern model-based fault diagnosis for non-linear systems:

*Solutions to Problem 1*

- Application of the unknown input observer for linear stochastic systems to form an Extended Unknown Input Observer (EUIO) (Section 4.1) for non-linear deterministic systems. A comprehensive convergence analysis with the Lyapunov approach was performed, resulting in convergence conditions for the EUIO. The obtained convergence conditions were utilised to improve the convergence properties of the EUIO.
- Development of an alternative EUIO structure (Section 4.2). A comprehensive convergence analysis was performed under less restrictive assumptions than those imposed in Section 4.1. The resulting convergence condition was employed to improve the convergence properties of the EUIO. It was also empirically verified that the EUIO of Section 4.2 is superior to the one of Section 4.1.

*Solution to Problem 2*

- Development of observers for Lipschitz non-linear systems (Section 4.3). Three different convergence criteria were developed with the Lyapunov method. Based on the achieved results, three different design procedures were proposed. These procedures were developed in such a way that the design problem was boiled down to solving a set of linear matrix inequalities or solving the generalised eigenvalue minimisation problem under LMI constraints, respectively.

*Solution to Problem 3*

- Adaptation of the genetic programming technique to discrete-time model construction and, especially, introduction of parameterised trees together with rules reducing an excessive number of parameters. In particular, effective genetic programming-based algorithms for designing both input-output and state-space models were developed. It was proven that the state-space models resulting from the above algorithms are asymptotically stable.

*Solutions to Problem 4*

- Application of the genetic programming technique to designing the EUIO. The problem of observer design was formulated as a global structure and parameter determination task with respect to instrumental matrices. To tackle the instrumental matrices selection problem, a genetic programming-based approach was proposed in Section 6.2.

- Application of the ESSS algorithm to designing the EUIO. To improve the convergence of the EUIO described in Section 4.2, the stochastic robustness technique was utilised to form a stochastic robustness metric describing unacceptable performance of the EUIO. In particular, it was shown that observer performance can be significantly improved with an appropriate selection of the instrumental matrix $\boldsymbol{Q}_k$. For that purpose, the B-spline approximation technique and evolutionary algorithms were utilised (Section 6.3).

These two approaches should also be perceived as solutions to the *Problem 1*. Indeed, apart from integrating evolutionary algorithms with extended unknown input observers, they significantly improve the convergence of such linearisation-based observers.

*Solutions to Problem 5*

- Development of robust neural network-based fault detection scheme with a multi-layer perceptron (Section 7.1). It was shown how to describe model uncertainty of the MLP with statistical techniques. Subsequently, two algorithms that can be used for decreasing such a model uncertainty with the use of experimental design theory were presented and described in detail. It was also shown how to use the resulting knowledge about model uncertainty for robust fault detection with the so-called adaptive threshold.
- Development of a robust neural network-based fault detection scheme with GMDH neural networks (Section 7.2). Starting from a set of input-output measurements of the system, it was shown how to estimate the parameters and the corresponding uncertainty of a neuron via the BEA. Subsequently, a complete design procedure of a neural network was proposed and carefully described. Based on the GMDH neural network, a novel robust fault detection scheme was proposed which supports diagnostic decisions.

*Solutions to Problem 6*

- It was revealed (Chapter 5) that appropriate scheduling of input signals significantly increases the performance of fault diagnosis through more accurate parameter estimation. It was also shown that active fault diagnosis for non-linear systems causes many difficult design problems. To tackle such challenging problems, it was proposed to employ experimental design theory. To show the effectiveness of such an approach, a complete development study was presented in Section 5.2. An experimental design theory-based approach to increasing the reliability of neural network-based FDI was also proposed and carefully described in Section 7.1.

The book also presents a number of practical implementations of the proposed approaches, which can be summarised as follows:

- Genetic programming-based design of an observer for a two-phase induction motor;
- ESSS algorithm-based design of an observer for a two-phase induction motor;

- State estimation and sensor, actuator fault diagnosis of a two-phase induction motor;
- State estimation of a one-link manipulator with revolute joints actuated by a DC motor;
- Experimental design for impedance measurement;
- Estimation and fault diagnosis of an impedance;
- State-space model design of a valve actuator with genetic programming;
- EUIO-based robust fault detection of a valve actuator;
- Experimental design for a neural model of a valve actuator;
- MLP-model design of a valve actuator;
- MLP-based robust fault detection of a valve actuator;
- GMDH neural network-based model design of a valve actuator;
- GMDH neural network-based robust fault detection of a valve actuator.

The advantage of general framework presented in this monograph is the fact that it is independent of a particular form of the system being diagnosed. Indeed, when the non-linear state-space model is available, then effective observer-based approaches can be employed. If this is not the case, then one can design such models with the proposed genetic programming approach. An alternative solution is to use the proposed robust neural network-based techniques. Finally, when the fault can be expressed in the form of changes in the physical parameters of the system, then parameter estimation-based strategies can be employed, as was the case for the impedance measurement problem.

Irrespective of the above advantage, there still remain open problems regarding some important design issues. What follows is a discussion of the areas proposed for further investigations.

*Integration of fault diagnosis and fault-tolerant control.* Most fault diagnosis techniques are developed as diagnostic or monitoring tools rather than an integral part of FTC systems. In other words, some existing FDI methods may not satisfy the requirements of controller reconfiguration. On the other hand, most reconfigurable controls are carried out under the assumption of possessing perfect information from the fault diagnosis system. To settle such challenging tasks in an efficient way, the following preliminary problems have to be solved:

- Providing clear and concise requirements regarding the fault diagnosis system in the FTC scheme,
- Providing analysis tools regarding interactions between the fault diagnosis system and reconfigurable control in an integrated framework.

*Development of active fault diagnosis for non-linear systems.* Most FDI schemes do not employ an additional design freedom that can be achieved by suitably scheduling input signals. Although there is a number of approaches that can be employed to solve this problem for linear system, there is no systematic approach that can be used for a general class of non-linear systems.

*Experimental design for dynamic neural networks.* It was clearly demonstrated in Section 7.1 that the application of experimental design to static neural networks results in neural models with a significantly smaller model uncertainty than those designed without it. This makes the resulting fault diagnosis system more reliable and effective. In spite of the incontestable appeal of such an approach, there are no effective experimental strategies that can be applied to dynamic neural networks. Indeed, the main limitation regarding the application of locally D-optimum on-line sequential design (Section 7.1.3) is the fact that the computation of D-optimal inputs involves an on-line global optimisation problem. Unfortunately, this task seems to be very difficult to solve on-line, which is required for dynamic neural networks.

*Relaxing the existence conditions of observers for Lipschitz systems.* As has been mentioned, many non-linear systems satisfy Lipschitz conditions. On the other hand, the Lipschitz constant for such systems may have relatively large values. This can make the usage of the observer design procedures described in Section 4.3 impossible. Apart from the simplicity of these procedures, the main reason for such a situation is the fact that the procedures are based on conservative transformation techniques. Thus, the development of less conservative transformation techniques constitutes one of the future research directions.

# Appendix

The main objective of this appendix is to show implementation details regarding the design procedures described in Section 4.3.2. Since the design procedures are very similar, the attention is restricted to the implementation of one of them, namely, the approach described in Section 4.3.2. The procedures presented in Sections 4.3.2 and 4.3.2 can easily be implemented by a minor modification of the the source code presented in this appendix.

In particular, the selection of $\boldsymbol{K}$ maximizing $\gamma$ for which the observer (4.136) is convergent is considered. Figure 8.1 presents the complete MATLAB source code that can be used to solve such a problem. As was mentioned in Section 4.3.2, the positivity of the right hand side of (4.176) is required for the well-posedness of the task and the applicability of polynomial-time interior point methods [62]. To tackle this problem, a simple remedy described in [62, pp. 8-41] is employed. As a result, instead of using (4.176) the following two LMIs are utilised:

$$\begin{bmatrix} \boldsymbol{X}-\boldsymbol{P} & \boldsymbol{C}^T\boldsymbol{L}^T-\boldsymbol{A}^T\boldsymbol{P} \\ \boldsymbol{L}\boldsymbol{C}-\boldsymbol{P}\boldsymbol{A} & -\boldsymbol{P} \end{bmatrix} \prec \begin{bmatrix} \boldsymbol{Y} & \boldsymbol{0} \\ \boldsymbol{0} & \boldsymbol{0} \end{bmatrix}, \tag{8.1}$$

and

$$\boldsymbol{Y} \prec \lambda(\beta+1)\boldsymbol{I}, \tag{8.2}$$

where $\boldsymbol{Y}$ is a symmetric matrix.

Now, let us describe the function presented in Fig. 8.1. The input and output parameters are $\boldsymbol{A}$, $\boldsymbol{C}$ and $\gamma$, $\boldsymbol{K}$, respectively. In the lines 2–4, all variables required in the LMIs (4.164), (4.174), (8.1) and (8.2) are set. The LMIs (4.164) are defined in the lines 5–7, while the lines 8–9 implement (4.174). Finally, the LMIs (8.1) and (8.2) are defined in the lines 10–14. Since all LMIs are defined, it is possible to find $\boldsymbol{K}$ maximizing $\gamma$ for which the observer (4.136) is convergent. As was described in Section 4.3.2, such an optimisation task is formulated as a generalised eigenvalue minimisation problem [20, 62]. The code of the lines 15–16 is employed to solve this problem. In the line 17, the correctness of the achieved solution is checked. Finally, in the lines 18–19 the maximum $\gamma$ for which the observer (4.136) is convergent and the corresponding $\boldsymbol{K}$ are calculated.

```
function [gamma,K]=GetGamma2(A,C)
 n=size(A,1); m=size(C,1); setlmis([]);
 P=lmivar(1,[n 1]); L=lmivar(2,[n m]); X=lmivar(1,[n 1]);
 Y=lmivar(1,[n 1]); Beta=lmivar(1,[1 1]);
 lmiterm([-1 1 1 P],1,1); lmiterm([-2 1 1 Beta],1,1);
 lmiterm([-3 1 1 Beta],1,1);lmiterm([-3 2 2 Beta],1,1);
 lmiterm([-3 2 1 P],1,1);
 lmiterm([-4 1 1 X],1,1); lmiterm([-4 2 2 0],1);
 lmiterm([-4 2 1 P],1,A); lmiterm([-4 2 1 L],-1,C);
 lmiterm([5 1 1 P],-1,1); lmiterm([5 2 2 P],-1,1);
 lmiterm([5 2 1 P],-1,A); lmiterm([5 2 1 L],1,C);
 lmiterm([5 1 1 Y],-1,1); lmiterm([5 1 1 X],1,1);
 lmiterm([6 1 1 Y],1,1); lmiterm([-6 1 1 Beta],1,1);
 lmiterm([-6 1 1 0],1);
 LMIss=getlmis;
 [lambda, popt]=gevp(LMIss,1);
 if lambda>0 error('The problem cannot be solved'); end;
 gamma=sqrt(-lambda); P=dec2mat(LMIss,popt,P);
 L=dec2mat(LMIss,popt,L); K=inv(P)*L;
```

**Fig. 8.1.** MATLAB® implementation of the design procedure of Section 4.3.2

# References


1. C. Aboky, G. Sallet, and J. C. Vivalda. Observers for Lipschitz nonlinear systems. *International Journal of Control*, 75(3):204–212, 2002.
2. E. Alcorta Garcia and P. M. Frank. Deterministic nonlinear observer-based approaches to fault diagnosis. *Control Engineering Practice*, 5(5):663–670, 1997.
3. A. Alessandri, T. Parisini, and R. Zoppoli. Neural approximators for non-linear finite-memory state estimation. *International Journal of Control*, 67(2):275–302, 1997.
4. B. D. O. Anderson and J. B. Moore. *Optimal Filtering.* Prentice-Hall, New Jersey, 1979.
5. L. Angrisani, A. Baccigalupi, and A. Pietrosanto. A digital signal-processing instrument for impedance measurement. *IEEE Transactions on Instrumentation and Measurement*, 45(6):930–934, 1996.
6. S. A. Ashton, D. N. Shields, and S. Daley. Design of a robust fault detection observer for polynomial non-linearities. In *Proc. 14th IFAC World Congress*, Beijing, P. R. China, 1999. CD-ROM.
7. A. C. Atkinson and A. N. Donev. *Optimum Experimental Designs.* Oxford University Press, New York, 1992.
8. S. S. Awad, N. Narasimhamurthi, and W. H. Ward. Analysis, design, and implementation of an AC bridge for impedance measurements. *IEEE Transactions on Instrumentation and Measurement*, 43(6):894–899, 1994.
9. M. Ayoubi. Fault diagnosis with dynamic neural structure and application to a turbo-charger. In *Proc. 2nd IFAC Symp. Fault Detection, Supervision and Safety of Technical Processes, SAFEPROCESS'94*, volume 2, pages 618–623, Espoo, Finland, 1994.
10. A. D. Back and A. C. Tsoi. FIR and IIR synapses. A new neural network architecture for time series modelling. *Neural Computation*, 3(3):375–385, 1991.
11. T. Back, U. Hammel, and H. P. Schwefel. Evolutionary computation: comments on the history and current state. *IEEE Transactions on Evolutionary Computation*, 1(1):3–17, 1997.
12. M. Bartyś, R. Patton, M. Syfert, S. Heras, and J. Quevedo. Introduction to the DAMADICS actuator FDI study. *Control Engineering Practice*, 14(6):577–596, 2006.
13. M. Basseville and I. V. Nikiforov. *Detection of Abrupt Changes: Theory and Application.* Prentice Hall, New York, 1993.

14. S. P. Bhattacharyya. Observer design for linear systems with unknown inputs. *IEEE Transactions on Automatic Control*, 23(3):483–484, 1978.
15. M. Blanke, S. Bogh, R. B. Jorgensen, and R. J. Patton. Fault detection for diesel engine actuator – a benchmark for FDI. In *Proc. 2nd IFAC Symp. Fault Detection, Supervision and Safety of Technical Processes, SAFEPROCESS'94*, volume 2, pages 498–506, Espoo, Finland, 1994.
16. M. Blanke, M. Kinnaert, J. Lunze, and M. Staroswiecki. *Diagnosis and Fault-Tolerant Control.* Springer-Verlag, New York, 2003.
17. E. P. J. Boer and E. M. T. Hendrix. Global optimization problems in optimal design of experiments in regression models. *Journal of Global Optimization*, 18(2):385–398, 2000.
18. M. Boutayeb. Identification of non-linear systemn in the presence of unknown but bounded disturbances. *IEEE Transactions on Automatic Control*, 45(8):1503–1507, 2000.
19. M. Boutayeb and D. Aubry. A strong tracking extended Kalman observer for nonlinear discrete-time systems. *IEEE Transactions on Automatic Control*, 44(8):1550–1556, 1999.
20. S. Boyd, E. Feron, L. E. Ghaoui, and V. Balakrishnan. *Linear Matrix Inequalities in System and Control Theory.* Siam, Philadelphia, 1994.
21. Z. Bubnicki. General approach to stability and stabilization for a class of uncertain discrete non-linear systems. *International Journal of Control*, 73(14):1298–1306, 2000.
22. K. Busawon, M. Saif, and M. Farza. A discrete-time observer for a class of non-linear systems. In *Proc. 36th IEEE Conference on Decision and Control, CDC*, pages 4796–4801, 1997.
23. C. Califano, S. Monaco, and D. Normand-Cyrot. On the observer design in discrete-time. *System and Control Letters*, 49(4):255–266, 2003.
24. S. L. Campbell and R. Nikoukhah. *Auxiliary Signal Design for Failure Detection.* Princeton University Press, Princeton, 2004.
25. P. Chandra and Y. Sing. Feedforward sigmoidal networks – equicontinuity and fault tolerance. *IEEE Transactions on Neural Networks*, 15(6):1350–1366, 2004.
26. H. G. Chen and K. W. Han. Improved quantitative measures of robustness for multivariable systems. *IEEE Transactions on Automatic Control*, 39(4):807–810, 1994.
27. J. Chen and R. J. Patton. *Robust Model Based Fault Diagnosis for Dynamic Systems.* Kluwer Academic Publishers, London, 1999.
28. J. Chen, R. J. Patton, and G. P. Liu. Optimal residual design for fault diagnosis using multi-objective optimization and genetic algorithms. *International Journal of Systems Science*, 27(6):567–576, 1996.
29. J. Chen, R. J. Patton, and H. Zhang. Design of unknown input observers and fault detection filters. *International Journal of Control*, 63(1):85–105, 1996.
30. Z. Chen, Y. He, F. Chu, and J. Huang. Evolutionary strategy for classification problems and its application in fault diagnosis. *Engineering Applications of Artificial Intelligence*, 16(1):31–38, 2003.
31. E. Y. Chow and A. A. Willsky. Analytical redundancy and the design of robust detection systems. *IEEE Transactions on Automatic Control*, 29(7):603–614, 1984.
32. G. Chryssolouris, M. Lee, and A. Ramsey. Confidence interval prediction for neural network models. *IEEE Transactions on Neural Networks*, 7(1):229–232, 1996.

33. T. Clement and S. Gentil. Reformulation of parameter identification with unknown-but-bounded errors. *Mathematics and Computers in Simulation*, 30(3):257–270, 1988.
34. T. Clement and S. Gentil. Recursive membership set estimation for output-error models. *Mathematics and Computers in Simulation*, 32(5–6):505–513, 1990.
35. D. A. Cohn. Neural network exploration using optimal experimental design. In *Advances in Neural Information Processing Systems 6 (Cowan, J. et al., (Eds.))*. Morgan Kaufman, San Mateo, 1994.
36. DAMADICS. Website of the research training network on Development and application of methods for actuator diagnosis in industrial control systems http://diag.mchtr.pw.edu.pl/damadics, 2004.
37. M. Darouach and M. Zasadzinski. Unbiased minimum variance estimation for systems with unknown exogenous inputs. *Automatica*, 33(4), 1997.
38. G. Dascupta and A. Huang. Assymptotically convergent modified least-squares with data dependent updating and forgetting factor for systems with bounded noise. *IEEE Transactions on Information Theory*, 33(3):383–391, 1987.
39. C. de Boor. *A Practical Guide to Splines*. Springer-Verlag, New York, 1978.
40. F. Delebecque, R. Nikoukah, and H. Rubio Scola. Test signal design for failure detection: A linear programming approach. *International Journal of Applied Mathematics and Computer Science*, 13(4):515–526, 2003.
41. I. S. Dhillon and S. Sra. *Modeling Data Using Directional Distributions*. Technical Report $TR - 03 - 06$, University of Texas of Austin, Austin, 2003.
42. S. X. Ding, Guo L., and Jeinsch T. A characterization of parity space and its application to robust fault detection. *IEEE Trans. Automatic Control*, 44(2):337–343, 1999.
43. W. Duch, J. Korbicz, L. Rutkowski, and R. Tadeusiewicz. *Biocybernetics and Biomedical Engineeering 2000, Neural Networks.*, volume 6. Academic Publishing Office EXIT, Warsaw, 2000.
44. M. Dutta, A. Rakshid, N. S. Battacharyya, and J. K. Choudhury. An application of the LMS adaptive algorithm for a digital AC bridge. *IEEE Transactions on Instrumentation and Measurement*, 36(2):894–897, 1987.
45. M. Dutta, A. Rakshid, and S.N. Battacharyya. Development and study of an automatic AC bridge for impedance measurement. *IEEE Transactions on Instrumentation and Measurement*, 50(5), 2001.
46. D. Erdogmus, O. Fontenela-Romero, and J. C. Principe. Linear-least-square initialization of multilayer perceptrons through backpropagation of the desired response. *IEEE Transactions on Neural Networks*, 16(2):325–337, 2005.
47. A. I. Esparcia-Alcazar. *Genetic Programming for Adaptive Digital Signal Processing*. PhD thesis, Glasgow University, 1998.
48. P. Fasconi, M. Gori, and G. Soda. Local feedback multilayered networks. *Neural Computation*, 4(1):120–130, 1992.
49. V. V. Fedorov. *Theory of Optimal Experiments*. Academic Press, New York, 1972.
50. V. V. Fedorov and P. Hackl. *Model-oriented Design of Experiments*. Springer, New York, 1997.
51. P. J Fleming and R. C. Purshouse. Evolutionary algorithms in control systems engineering: a survey. *Control Engineering Practice*, 10(11):1223–1241, 2002.
52. R. Fletcher. *Practical Methods of Optimization*. John Wiley and Sons, Chichester, New York, Brisbane, Toronto, 1981.
53. D. B. Fogel. *Evolutionary Computation: Toward a New Philosophy of Machine Intelligence*. Willey-IEEE Press, New York, 1995.

54. L. J. Fogel, A. J. Owens, and M. J. Walsh. An overview of evolutionary programming. In *Evolutionary Algorithms (Davis L. D., De Jong M. D., Vose M. D. and Whitley L. D. (Eds.)*, pages 89–109. Springer-Verlag, Heidelberg, 1999.
55. I. Ford, D. M. Titterington, and C. P. Kitsos. Recent advances in nonlinear experimental design. *Technometrics*, 31(1):49–60, 1989.
56. P. M. Frank and S. X. Ding. Survey of robust residual generation and evaluation methods in observer-based fault detection systems. *Journal of Process Control*, 7(6):403–424, 1997.
57. P. M. Frank, G. Schreier, and E. A. Garcia. Nonlinear observers for fault detection and isolation. In *New Directions in Nonlinear Observer Design (H.Nijmeijer, T.Fossen (Eds.))*. Springer-Verlag, Berlin, 1999.
58. M. J. Fuente and S. Saludes. Fault detection and isolation in a non-linear plant via neural networks. In *Proc. 4th IFAC Symp. Fault Detection, Supervision and Safety for Technical Processes, SAFEPROCESS 2000*, volume 1, pages 472–477, Budapest, Hungary, 2000.
59. K. Fukumizu. Active learning in multilayer perceptrons. In *Advances in Neural Information Processing Systems (Touretzky D. S. (Ed.))*, pages 295–301. MIT Press, Cambridge, 1996.
60. K. Fukumizu. A regularity condition of the information matrix of a multi-layer perceptron network. *Neural Networks*, 9(5):871–879, 1996.
61. K. Fukumizu. Statistical active learning in multilayer perceptrons. *IEEE Transactions on Neural Networks*, 11(1):17–26, 2000.
62. P. Gahinet, A. Nemirovski, A. J. Laub, and M. Chilali. *LMI Control Toolbox. For Use With Matlab.* The MathWorks Inc., Natick, MA, 1995.
63. R. Galar. Evolutionary search with soft selection. *Biological Cybernetics*, 60(1):124–141, 1989.
64. X. Z. Gao and S. J. Ovaska. Motor fault detection and diagnosis using soft computing method. In *Soft Computing in Industrial Electronics (S. J. Ovaska (Ed.))*, pages 3–45. Springer, Heidelberg, 2002.
65. J. Gertler. *Fault Detection and Diagnosis in Engineering Systems.* Marcel Dekker, New York, 1998.
66. G. C. Goodwin and R. L. Payne. *Dynamic Systems Identification. Experiment Design and Data Analysis.* Acadamic Press, New York, 1977.
67. G. C. Goodwin and K. S. Sin. *Adaptive Filtering, Prediction and Control.* Prentice-Hall, New Jersey, 1984.
68. M. Gori, Y. Bengio, and R. D. Mori. BPS: A learning algorithm for capturing the dynamic nature of speech. In *Proc. Int. Joint Conf. Neural Networks*, volume II, pages 417–423, 1989.
69. G. J. Gray, D. J. Murray-Smith, Y. Li, K. C. Sharman, and T. Weinbrenner. Non-linear model structure identification using genetic programming. *Control Engineering Practice*, 6(11):1341–1352, 1998.
70. C. Guernez, J. Ph. Cassar, and M. Staroswiecki. Extension of parity space to non-linear polynomial dynamic systems. In *Proc. 3rd IFAC Symp. Fault Detection, Supervision and Safety of Technical Processes, SAFEPROCESS'97*, volume 2, pages 861–866, Hull, UK, 1997.
71. G. Guglielmi, T. Parisini, and G. Rossi. Fault diagnosis and neural networks: A power plant application. *Control Engineering Practice*, 3(5):601–620, 1995.
72. L. Z. Guo and Q. M. Zhu. A fast convergent extended Kalman observer for non-linear discrete-time systems. *International Journal of Systems Science*, 33(13):1051–1058, 2002.

73. M. M. Gupta, L. Jin, and N. Homma. *Static and Dynamic Neural Networks. From Fundamentals to Advanced Theory.* Wiley, New Jersey, 2003.
74. M. Gutowski. Lévy flights as an underlying mechanism for a global optimization algorithm. In *Proc. 5th Conf. Evolutionary Algorithms and Global Optimization, Warsaw University of Technology Press*, pages 79–86, Jastrzębia Góra, Poland, 2001.
75. A. Hac. Design of disturbance decoupled observers for bilinear systems. *ASME Journal of Dynamic Systems, Measurement, and Control*, 114:556–562, 1992.
76. J. Hertz, R. Krogh, and G. Palmer. *Introduction to the Neural Computation.* Addison-Wesley Publishing Company Inc., New York, 1991.
77. J. H. Holland. *Adaptation in natural and artificial systems.* The University of Michigan Press, Ann Arbor, MI, 1975.
78. K. Hornik, M. Stinchombe, and H. White. Multi-layers feed-forward networks are universal approximators. *Neural Computation*, 2:359–366, 1989.
79. M. Hou and P. C. Mueller. Design of observers for linear systems with unknown inputs. *IEEE Trans. Automatic Control*, 37(6):871–875, 1992.
80. M. Hou and R.J. Patton. Optimal filtering for systems with unknown inputs. *IEEE Trans. Automatic Control*, 43(3):445–449, 1998.
81. M. Hou and A. C. Pugh. Observing state in bilinear systems: an UIO approach. In *Proc. IFAC Symp. Fault Detection, Supervision and Safety of Technical Processes: SAFEPROCESS'97*, volume 2, pages 783–788, Hull, UK, 1997.
82. S. Hui and S.H. Zak. Observer design for systems with unknown input. *International Journal of Applied Mathematics and Computer Science*, 15(4):431–446, 2005.
83. K. J. Hunt, D. Sbarbaro, R. Zbikowski, and P. J. Gawthrop. Neural networks for control systems - a survey. *Automatica*, 28(6):1083–1112, 1992.
84. R. Iserman. *Fault diagnosis systems. An introduction from fault detection to fault tolerance.* Springer-Verlag, New York, 2006.
85. A. G. Ivakhnenko and J. A. Mueller. Self-organizing of nets of active neurons. *System Analysis Modelling Simulation*, 20:93–106, 1995.
86. A. Janczak. *Identification of Nonlinear Systems Using Neural Networks and Polynomial Models. A Block-oriented Approach.* Lecture Notes in Control and Information Sciences. Springer–Verlag, Berlin, 2005. M. Thoma, M. Morari (Eds.).
87. J. Kaczmarek, R. Rybski, and D. Uciński. A recursive DSP approach to impedance measurement. In *Proc. Symp. Development in Digital Measuring Instrumentation, IMEKO TC-4*, Naples, Italy, 1998.
88. I. Karcz-Dulęba. Asymptotic behaviour of a discrete dynamical system generated by a simple evolutionary process. *International Journal of Applied Mathematics and Computer Science*, 14(1):79–90, 2004.
89. M. Karpenko, N. Sepehri, and D. Scuse. Diagnosis of process valve actuator faults using a multilayer neural network. *Control Engineering Practice*, 11(11):1289–1299, 2003.
90. M. P. Kazimierkowski and T. Orłowska-Kowalska. Neural network estimation and N-F control in converter-fed induction motor drives. In *Soft Computing in Industrial Electronics (S. J. Ovaska (Ed.))*, pages 45–95. Springer, Heidelberg, 2002.
91. J. Y. Keller and M. Darouach. Two-stage Kalman estimator with unknown exogenous inputs. *Automatica*, 35(2), 1999.
92. M. Kinneart. Robust fault detection based on observers for bilinear systems. *Automatica*, 35(11):1829–1842, 1999.

93. N. Kobayashi and Nakamizo R. An observer design for linear systems with unknown inputs. *International Journal of Control*, 17(3):471–479, 1982.
94. D. Koenig and S. Mammar. Desing of a class of reduced unknown inputs nonlinear observer for fault diagnosis. In *Proc. American Control Conference, ACC*, Arlington, USA, 2002.
95. J. Korbicz. Fault detection using analytical and soft computing methods. *Bulletin of the Polish Academy of Sciences: Technical Sciences*, 54(1):75–88, 2006.
96. J. Korbicz, J. Kościelny, Z. Kowalczuk, and W. Cholewa (Eds.). *Fault diagnosis. Models, Artificial Intelligence, Applications.* Springer-Verlag, Berlin, 2004.
97. J. Korbicz (Ed.). Special section on soft computing in control and fault diagnosis. *International Journal of Applied Mathematics and Computer Science*, 16(1):7–115, 2006.
98. J. M. Kościelny. *Diagnostics of Automatic Industrial Processes.* Academic Publishing Office EXIT, Warsaw, 2001.
99. M. Kowal. *Optimization of neuro-fuzy structures in technical diagnostics.* Lecture Notes in Control and Computer Science, Vol. 9. University of Zielona Góra Press, Zielona Góra, 2005.
100. Z. Kowalczuk and K. Gunawickrama. Leak detection and isolation for transission pipelines via non-linear state estimation. In *Proc. 4th IFAC Symp., Fault Detection, Supervision and Safety of Technical Processes, SAFEPROCESS 2000*, volume 2, pages 943–948, Budapest, Hungary, 2000.
101. Z. Kowalczuk and P. Suchomski. Analytical redundancy: Kalman filter as a basis for robust detection. In *Proc. 5th Nat. Conf. Diagnostics of Industrial Processes, DPP 2001*, Lagów Lubuski, Poland, 2001.
102. Z. Kowalczuk, P. Suchomski, and T. Bialaszewski. Evolutionary muti-objective pareto optimization of diagnostic state observers. *International Journal of Applied Mathematics and Computer Science*, 9(3):689–709, 1999.
103. J. R. Koza. *Genetic Programming: On the Programming of Computers by Means of Natural Selection.* The MIT Press, Cambridge, 1992.
104. V. Krishnaswami and G. Rizzoni. Non-linear parity equation residual generation for fault detection and isolation. In *Proc. 2nd IFAC Symp. Fault Detection, Supervision and Safety of Technical Processes, SAFEPROCESS'94*, volume 1, pages 317–322, Espoo, Finland, 1994.
105. T. T. Le, J. Watton, and D. T. Pham. An artificial neural network-based approach to fault diagnosis and classification of fluid power systems. *Journal of Systems and Control Engineering*, 211(4):307–317, 1997.
106. T. T. Le, J. Watton, and D. T. Pham. Fault classification of fluid power systems using a dynamics feature extraction technique and neural networks. *Journal of Systems and Control Engineering*, 212(2):87–97, 1998.
107. Ch.Y. Lee and X. Yao. Evolutionary programming using mutations based on the Lévy probability distribution. 8(1):1–13, 2004.
108. D. MacKay. Information-based objective functions for active data selection. *Neural Computation*, 4(4):305–318, 1992.
109. D. Maksarow and J. Norton. State bounding with ellipsoidal set description of the uncertainty. *International Journal of Control*, 65(5):847–866, 1996.
110. T. Marcu. A multiobjective evolutionary approach to pattern recognition for robust diagnosis of process faults. In *Proc. 3rd IFAC Symp. Fault detection, supervision and safety for technical processes, SAFEPROCESS'97*, pages 1183–1188, Hull, UK, 1997.
111. K. V. Mardia and P. Jupp. *Directional Statistics.* John Willey and Sons, New York, 2000.

112. Ch. I. Marrison and R. F. Stengel. Robust control system design using random search and genetic algorithms. *IEEE Transactions on Automatic Control*, 42(6):835–839, 1997.
113. B. K. Massoumnia and W. E. Vander Velde. Generating parity relations for detecting and identifying control system components failures. *Journal of Guidance, Control and Dynamics*, 11(1):60–65, 1988.
114. M. Metenidis, M. Witczak, and J. Korbicz. A novel genetic programming approach to nonlinear system modelling: application to the DAMADICS benchmark problem. *Engineering Applications of Artificial Intelligence*, 17(4):363–370, 2004.
115. Z. Michalewicz. *Genetic Algorithms + Data Structures = Evolution Programs*. Springer, Berlin, 1996.
116. M. Milanese, J. Norton, H. Piet-Lahanier, and E. Walter. *Bounding Approaches to System Identification*. Plenum Press, New York, 1996.
117. J. A. Miller, W. D. Potter, R. V. Gandham, and C. N. Lapena. An evaluation of local improvement operators for genetic algorithms. *IEEE Transactions on Systems, Man and Cybernetics*, 23(5):1340–1351, 1993.
118. S. H. Mo and J. P. Norton. Fast and robust algorithm to compute exact polytope parameter bounds. *Mathematics and Computers in Simulation*, 32:481–493, 1990.
119. M. Mrugalski. *Neural Network Based Modelling of Non-linear Systems in Fault Detection Schemes*. PhD thesis, Faculty of Electrical Engineering, Computer Science and Telecommunications, University of Zielona Góra, Zielona Góra, 2004.
120. M. Mrugalski and M. Witczak. Parameter estimation of dynamic GMDH neural networks with the bounded-error technique. *Journal of Applied Computer Science*, 10(1):77–90, 2002.
121. J. E. Mueller and F. Lemke. *Self-organising Data Mining*. Libri, Hamburg, 2000.
122. K. S. Narendra and K. Parthasarathy. Identification and control of dynamical systems using neural networks. *IEEE Transactions on Neural Networks*, 1(1):12–18, 1990.
123. O. Nelles. *Non-linear Systems Identification. From Classical Approaches to Neural Networks and Fuzzy Models*. Springer, Berlin, 2001.
124. R. Nikoukhah, S. L. Campbell, and F. Delebecque. Detection signal design for failure detection: a robust approach. *International Journal of Adaptive Control and Signal Processing*, (14):701–724, 2000.
125. A. Obuchowicz. *Evolutionary Algorithms in Global Optimization and Dynamic System Diagnosis. ISBN: 83-88317-02-4*. Monographs, 3. Lubuskie Scientific Society, Zielona Góra, 2003.
126. A. Obuchowicz and J. Korbicz. Evolutionary methods in designing diagnostic systems. In *Fault Diagnosis. Models, Artificial Intelligence, Applications. (Korbicz J., Kościelny J.M., Kowalczuk Z., and Cholewa W. (Eds.))*, pages 301–331. Springer-Verlag, Berlin, 2004.
127. A. Obuchowicz and P. Prętki. Phenotypic evolution with mutation based on symmetric $\alpha$-stable distribution. *International Journal of Applied Mathematics and Computer Science*, 14(3):289–316, 2004.
128. Papers of the special sessions. DAMADICS I, II, III. In *Proc. 5th IFAC Symp. Fault Detection Supervision and Safety of Technical Processes, SAFEPROCESS 2003*, Washington DC, USA, 2003. June 9-11.
129. T. Orłowska-Kowalska. *Sensorless driving systems with induction motors*. Technical University of Wrocław Press, Wrocław, 2003.
130. S. Osowski, J. Herault, and P. Demartines. Fault location in analogue circuits using Kohonen nerual network. *Bulletin of the Polish Academy of Sciences: Technical Sciences*, 43(1):111–123, 1995.

131. S. Osowski and R. Sałat. Fault location in transmission line using hybrid neural networks. *Compel*, 21(1):13–30, 2002.
132. S. J. Ovaska (Ed.). *Soft Computing in Industrial Electronics.* Springer, Heidelberg, 2002.
133. G. Papadopoulos, P. J. Edawrds, and A. F. Murray. Confidence estimation methods for neural networks: A practical comparison. *IEEE Transactions on Neural Networks*, 12(6):1279–1287, 2001.
134. P. N. Paraskevopoulos. *Digital Control Systems.* Prentice Hall, London, 1996.
135. T Parisini, Polycarpou M., Sanguineti M., and Vemuri A. Robust parametric and non-parametric fault diagnosis in nonlinear input-output systems. In *Proc. 36th Conf. Decision and Control*, pages 4481–4482, San Diego, USA, 1997.
136. K. Patan and J. Korbicz. Artificial neural networks in fault diagnosis. In *Fault Diagnosis. Models, Artificial Intelligence, Applications. (Korbicz J., Kościelny J.M., Kowalczuk Z., and Cholewa W. (Eds.))*, pages 333–379. Springer-Verlag, Berlin, 2004.
137. K. Patan and T. Parisini. Identification of neural dynamic models for fault detection and isolation: the case of a real sugar evaporation process. *Journal of Process Control*, 15(1):67–79, 2005.
138. R. J. Patton, P. M. Frank, and R. N. Clark. *Issues of Fault Diagnosis for Dynamic Systems.* Springer-Verlag, Berlin, 2000.
139. R. J. Patton and J. Korbicz. Special issue: Advances in computational intelligence for fault dignosis systems. *International Journal of Applied Mathematics and Computer Science*, 9(3), 1999.
140. R. J. Patton, J. Korbicz, M. Witczak, and F. J. Uppal. Combined computational intelligence and analytical methods in fault diagnosis. In *Intelligent Control Systems using Computational Intelligence Techniques, (Ruano, A. E. (Ed.))*, pages 349–385. The IEE Press, London, 2005.
141. Y. B. Peng, A. Youssouf, P. Arte, and M. Kinnaert. A complete procedure for residual generation and evaluation with application to a heat exchanger. *IEEE Transactions on Control Systems Technology*, 5(6):542–555, 1997.
142. A. M. Pertew, H. J. Marquez, and Q. Zhao. $\mathcal{H}_\infty$ synthesis of unknown input observers for non-linear lipschitz systems. *International Journal of Control*, 78(15):1155–1165, 2005.
143. D. T. Pham and L. Xing. *Neural Networks for Identification, Prediction and Control.* Springer-Verlag, London, 1995.
144. L. L. Porter and K. M. Passino. Genetic adaptive observers. *Engineering Applications of Artificial Intelligence*, 8(3):261–269, 1995.
145. P. Prętki and A. Obuchowicz. Directional distribution and their application to evolutionary algorithms. In *Lecture Notes in Artificial Intelligence: Artificial Intelligence and Soft Computing*, pages 440–449. Springer-Verlag, Berlin, 2006.
146. L. Pronzato and E. Walter. Sequential experimental design for parameter bounding. In *Proc. European Control Conference, ECC*, pages 1181–1186, Grenoble, 1991.
147. L. Pronzato and E. Walter. Eliminating suboptimal local minimizers in nonlinear parameter estimation. *Technometrics*, 43(4):434–442, 2001.
148. S. Qiang, X. Z. Gao, and X. Zhunag. State-of-the-art in soft computing-based motor fault diagnosis. In *Proc. IEE Int. Conf. Control Applications*, pages 1381–1386, Istanbul, Turkey, 2003.
149. R. Rajamani. Observers for Lipschitz non-linear systems. *IEEE Transactions on Automatic Control*, 43(3):397–401, 1998.

150. R. Rajamani and Y. M. Cho. Existence and design of observers for nonlinear systems: relation to distance to unobservability. *International Journal of Control*, 69(5):717–731, 1998.
151. A. E. Ruano (Ed.). *Intelligent Control Systems using Computational Intelligence Techniques.* The IEE Press, London, 2005.
152. N.T. Russell, H.H.C. Bakker, and R.I. Chaplin. Modular neural network modelling for long-range prediction of an evaporator. *Control Engineering Practice*, 8(1):49–59, 2000.
153. D. Rutkowska. *Neuro-Fuzzy Architectures and Hybrid Learning.* Springer, Berlin, 2002.
154. L. Rutkowski. *New Soft Computing Techniques for System Modelling, Patern Classification and Image Processing.* Springer, Berlin, 2004.
155. G. Schreier, J. Ragot, R. J. Patton, and P. M. Frank. Observer design for a class of non-linear systems. In *Proc. 3rd IFAC Symp. Fault Detection, Supervision and Safety of Technical Processes, SAFEPROCESS'97*, volume 1, pages 483–488, Hull, UK, 1997.
156. G. A. F. Seber and C. J. Wild. *Nonlinear Regression.* John Wiley and Sons, New York, 1989.
157. R. Seliger and P. Frank. Robust observer-based fault diagnosis in non-linear uncertain systems. In *Issues of Fault Diagnosis for Dynamic Systems (Patton ,Frank and Clark, (Eds.)).* Springer-Verlag, Berlin, 2000.
158. D. N. Shields and S. Ashton. A fault detection observer method for non-linear systems. In *Proc. 4th IFAC Symp. Fault Detection, Supervision and Safety of Technical Processes, SAFEPROCESS 2000*, volume 1, pages 226–231, Budapest, Hungary, 2000.
159. A. Shumsky. Robust residual generation for diagnosis of non-linear systems: parity relation approach. In *Proc. 3rd IFAC Symp. Fault Detection, Supervision and Safety of Technical Processes, SAFEPROCESS'97*, volume 2, pages 867–872, Hull, UK, 1997.
160. J. Sjoberg, Q. Zhang, L. Ljung, A. Benveniste, B. Delyon, P. Y. Glorennec, H. Hjalmarsson, and A. Juditsky. Non-linear black-box modeling in system identification: A unified overview. *Automatica*, 31(12):1691–1724, 1995.
161. T. Soderstrom and P. Stoica. *System Identification.* Prentice-Hall International, Hemel Hempstead, 1989.
162. R. Sun, F. Tsung, and L. Qu. Combining bootstrap and genetic programming for feature discovery in diesel engine diagnosis. *International Journal of Industrial Engineering*, 11(3):273–281, 2004.
163. P. Supavatanakul. Timed discrete-event approach to the diagnosis of the actuator benchmark. In *Proc. 4th DAMADICS Workshop on Qualitative Approach for Fault Diagnosis*, Bochum, Germany, 2002.
164. H. J. Sussmann. Uniqueness of the weights for minimal feedworward nets with a given input-output map. *Neural Networks*, 5:589–593, 1992.
165. R. Tadeusiewicz and Ogiela M. R. *Medical Image Understanding Technology. Artificial Intelligence and Soft Computing for Image Understanding.* Springer, Berlin, 2004.
166. F. E. Thau. Observing the state of nonlinear dynamic systems. *International Journal of Control*, 17(3):471–479, 1973.
167. D. Uciński. *Optimal Measurement Methods for Distributed Parameter System Identification.* Systems and Control Series. CRC Press, Boca Raton, 2005.

168. F.J. Uppal, R. J. Patton, and M. Witczak. A neuro-fuzzy multiple-model observer approach to robust fault diagnosis based on the DAMADICS benchmark problem. *Control Engineering Practice*, 14(6):699–717, 2006.
169. B.K. Walker and H. Kuang-Yang. FDI by extended Kalman filter parameter estimation for an industrial actuator benchmark. *Control Engineering Practice*, 3(12):1769–1774, 1995.
170. E. Walter and L. Pronzato. *Identification of Parametric Models from Experimental Data.* Springer, London, 1996.
171. S. H. Wang, E. J. Davison, and P. Dorato. Observing the states of systems with unmeasurable disturbances. *IEEE Transactions on Automatic Control*, 20(5):716–717, 1975.
172. Z. Wang and H. Unbehauen. A class of non-linear observers for discrete-time systems with parametric uncertainty. *International Journal of Systems Science*, 31(1):19–26, 2000.
173. J. Watkins and S. Yurkovich. Set-membership strategies for fault detection and isolation. In *Proc. IEEE–SMC Symposium on Modelling, Analysis and Simulation, CESA*, pages 824–830, Lille, France, 1996.
174. J. Watton and D. T. Pham. An artificial NN based approach to fault diagnosis and classification of fluid power systems. *Journal of System and Control Engineering*, 211(4):307–317, 1997.
175. M. Weerasinghe, J. B. Gomm, and D. Williams. Neural networks for fault diagnosis of a nuclear fuel processing plant at different operating points. *Control Engineering Practice*, 6(2):281–289, 1998.
176. H. White. Learning in artificial neural networks: A statistical perspective. *Neural Computation*, 4(4):305–318, 1989.
177. R. J. Williams and D. Zipser. A learning algorithm for continually running fully recurrent neural networks. *Neural Computation*, 1(2):270–280, 1989.
178. A. S. Willsky and H. L. Jones. A generalized likelihood ratio approach to the detection of jumps in linear systems. *IEEE Transactions on Automatic Control*, 21(1):108–112, 1976.
179. M. Witczak. *Identification and fault detection of non-linear dynamic systems.* Lecture Notes in Control and Computer Science, Vol. 1. University of Zielona Góra Press, Zielona Góra, 2003.
180. M. Witczak. Developing D-optimum experimental conditions for model-based fault detection systems. In *Proc. 16th IFAC World Congress*, Prague, Czech Republic, 2005.
181. M. Witczak. Advances in model-based fault diagnosis with evolutionary algorithms and neural networks. *International Journal of Applied Mathematics and Computer Science*, 16(1):85–99, 2006.
182. M. Witczak. Toward the training of feed-forward neural networks with the D-optimum input sequence. *IEEE Transactions on Neural Networks*, 17(2):357–373, 2006.
183. M. Witczak and J. Korbicz. Observers and genetic programming in the identification and fault diagnosis of non-linear dynamic systems. In *Fault Diagnosis. Models, Artificial Intelligence, Applications. (Korbicz J., Kościelny J. M., Kowalczuk Z., and Cholewa W. (Eds.))*, pages 457–509. Springer-Verlag, Berlin, 2004.
184. M. Witczak and J. Korbicz. Design of observers for Lipschitz non-linear discrete-time systems. In *Proc. 14th IFAC Symp. System Identification, SYSID 2006*, pages 985–990, Newcastle, Australia, 2006. + CD-ROM.

185. M. Witczak, J. Korbicz, M. Mrugalski, and R. J. Patton. A GMDH neural network-based approach to robust fault diagnosis: application to the DAMADICS benchmark problem. *Control Engineering Practice*, 14(6):671–683, 2006.
186. M. Witczak, J. Korbicz, and V. Puig. An LMI approach to designing observers and unknown input observers for nonlinear systems. In *Proc. 6th IFAC Symposium on Fault Detection Supervision and Safety of Technical Processes, SAFEPROCESS 2006*, Beijing, China, 2006. + CD-ROM.
187. M. Witczak, A. Obuchowicz, and J. Korbicz. Genetic programming based approaches to identification and fault diagnosis of non-linear dynamic systems. *International Journal of Control*, 75(13):1012–1031, 2002.
188. M. Witczak and P. Prętki. Designing neural-network-based fault detection systems with D-optimum experimental conditions. *Computer Assisted Mechanics and Engineering Sciences*, 12(2):279–291, 2005.
189. M. Witczak and P. Prętki. Design of an extended unknown input observer with stochastic robustness techniques and evolutionary algorithms. *International Journal of Control*, 2007. In print.
190. G. G. Yen and K. Lin. Wavelet packet feature extraction for vibration monitoring. *IEEE Transactions on Industrial Electronics*, 47(3):650–667, 2000.
191. D. Yu and D. N. Shileds. Bilinear fault detection observer and its application to a hydraulic system. *International Journal of Control*, 64:1023–1047, 1996.
192. W. Yu and L. Xiaoou. Some new results on system identification with dynamic neural networks. *IEEE Transactions on Neural Networks*, 12(2):412–417, 2001.
193. H. Y. Zhang, C. W. Chan, K. C. Cheung, and Hong J. Nonlinear adaptive observer design based on b-spline neural network. In *14-th IFAC World Congress*, pages 213–217, Beijing, P. R. China, 1999.
194. J. Zhang, J. Ma, and Y. Yan. Assessing blockage of the sensing line in a differential-pressure flow sensor by using the wavelet transform of its output. *Measurement Science and Technology*, 11(3):178–184, 2000.
195. J. Q. Zhang, S. J. Ovaska, and Z. Xinmin. Development and study of an automatic AC bridge for impedance measurement. *IEEE Transactions on Instrumentation Measurement*, 47(2), 1998.
196. X. J. Zhang. *Auxiliary Signal Design in Fault Detection and Diagnosis*. Springer-Verlag, Heidelberg, 1989.
197. J. Zhao, B. Chen, and J. Shen. Multidimensional nonorthogonal wavelet-sigmoid basis function neural networkfor dynamic process fault diagnosis. *Computers and Chemical Engineering*, 23(1):83–92, 1998.
198. A. Zolghadri, D. Henry, and M. Monsion. Design of nonlinear observers for fault diagnosis. a case study. *Control Engineering Practice*, 4(11):1535–1544, 1996.

# Index

Printing: Mercedes-Druck, Berlin
Binding: Stein+Lehmann, Berlin

# Lecture Notes in Control and Information Sciences

**Edited by M. Thoma, M. Morari**

Further volumes of this series can be found on our homepage:
springer.com

**Vol. 355:** Zhang H.; Xie L.:
Control and Estimation of Systems with Input/Output Delays
213 p. 2007 [978-3-540-71118-6]

**Vol. 354:** Witczak M.
Modelling and Estimation Strategies for Fault Diagnosis of Non-Linear Systems
208 p. 2007 [978-3-540-71114-8]

**Vol. 353:** Bonivento C.; Isidori A.; Marconi L.; Rossi C. (Eds.)
Advances in Control Theory and Applications
305 p. 2007 [978-3-540-70700-4]

**Vol. 352:** Chiasson, J.; Loiseau, J.J. (Eds.)
Applications of Time Delay Systems
358 p. 2007 [978-3-540-49555-0]

**Vol. 351:** Lin, C.; Wang, Q.-G.; Lee, T.H., He, Y.
LMI Approach to Analysis and Control of Takagi-Sugeno Fuzzy Systems with Time Delay
204 p. 2007 [978-3-540-49552-9]

**Vol. 350:** Bandyopadhyay, B.; Manjunath, T.C.; Umapathy, M.
Modeling, Control and Implementation of Smart Structures 250 p. 2007 [978-3-540-48393-9]

**Vol. 349:** Rogers, E.T.A.; Galkowski, K.; Owens, D.H.
Control Systems Theory and Applications for Linear Repetitive Processes 482 p. 2007 [978-3-540-42663-9]

**Vol. 347:** Assawinchaichote, W.; Nguang, K.S.; Shi P.
Fuzzy Control and Filter Design for Uncertain Fuzzy Systems
188 p. 2006 [978-3-540-37011-6]

**Vol. 346:** Tarbouriech, S.; Garcia, G.; Glattfelder, A.H. (Eds.)
Advanced Strategies in Control Systems with Input and Output Constraints
480 p. 2006 [978-3-540-37009-3]

**Vol. 345:** Huang, D.-S.; Li, K.; Irwin, G.W. (Eds.)
Intelligent Computing in Signal Processing and Pattern Recognition
1179 p. 2006 [978-3-540-37257-8]

**Vol. 344:** Huang, D.-S.; Li, K.; Irwin, G.W. (Eds.)
Intelligent Control and Automation
1121 p. 2006 [978-3-540-37255-4]

**Vol. 341:** Commault, C.; Marchand, N. (Eds.)
Positive Systems
448 p. 2006 [978-3-540-34771-2]

**Vol. 340:** Diehl, M.; Mombaur, K. (Eds.)
Fast Motions in Biomechanics and Robotics
500 p. 2006 [978-3-540-36118-3]

**Vol. 339:** Alamir, M.
Stabilization of Nonlinear Systems Using Receding-horizon Control Schemes
325 p. 2006 [978-1-84628-470-0]

**Vol. 338:** Tokarzewski, J.
Finite Zeros in Discrete Time Control Systems
325 p. 2006 [978-3-540-33464-4]

**Vol. 337:** Blom, H.; Lygeros, J. (Eds.)
Stochastic Hybrid Systems
395 p. 2006 [978-3-540-33466-8]

**Vol. 336:** Pettersen, K.Y.; Gravdahl, J.T.; Nijmeijer, H. (Eds.)
Group Coordination and Cooperative Control
310 p. 2006 [978-3-540-33468-2]

**Vol. 335:** Kozłowski, K. (Ed.)
Robot Motion and Control
424 p. 2006 [978-1-84628-404-5]

**Vol. 334:** Edwards, C.; Fossas Colet, E.; Fridman, L. (Eds.)
Advances in Variable Structure and Sliding Mode Control
504 p. 2006 [978-3-540-32800-1]

**Vol. 333:** Banavar, R.N.; Sankaranarayanan, V.
Switched Finite Time Control of a Class of Underactuated Systems
99 p. 2006 [978-3-540-32799-8]

**Vol. 332:** Xu, S.; Lam, J.
Robust Control and Filtering of Singular Systems
234 p. 2006 [978-3-540-32797-4]

**Vol. 331:** Antsaklis, P.J.; Tabuada, P. (Eds.)
Networked Embedded Sensing and Control
367 p. 2006 [978-3-540-32794-3]

**Vol. 330:** Koumoutsakos, P.; Mezic, I. (Eds.)
Control of Fluid Flow
200 p. 2006 [978-3-540-25140-8]

**Vol. 329:** Francis, B.A.; Smith, M.C.; Willems, J.C. (Eds.)
Control of Uncertain Systems: Modelling, Approximation, and Design
429 p. 2006 [978-3-540-31754-8]

**Vol. 328:** Loría, A.; Lamnabhi-Lagarrigue, F.; Panteley, E. (Eds.)
Advanced Topics in Control Systems Theory
305 p. 2006 [978-1-84628-313-0]

**Vol. 327:** Fournier, J.-D.; Grimm, J.; Leblond, J.; Partington, J.R. (Eds.)
Harmonic Analysis and Rational Approximation
301 p. 2006 [978-3-540-30922-2]

**Vol. 326:** Wang, H.-S.; Yung, C.-F.; Chang, F.-R.
$H_\infty$ Control for Nonlinear Descriptor Systems
164 p. 2006 [978-1-84628-289-8]

**Vol. 325:** Amato, F.
Robust Control of Linear Systems Subject to Uncertain
Time-Varying Parameters
180 p. 2006 [978-3-540-23950-5]

**Vol. 324:** Christofides, P.; El-Farra, N.
Control of Nonlinear and Hybrid Process Systems
446 p. 2005 [978-3-540-28456-7]

**Vol. 323:** Bandyopadhyay, B.; Janardhanan, S.
Discrete-time Sliding Mode Control
147 p. 2005 [978-3-540-28140-5]

**Vol. 322:** Meurer, T.; Graichen, K.; Gilles, E.D. (Eds.)
Control and Observer Design for Nonlinear Finite and Infinite Dimensional Systems
422 p. 2005 [978-3-540-27938-9]

**Vol. 321:** Dayawansa, W.P.; Lindquist, A.; Zhou, Y. (Eds.)
New Directions and Applications in Control Theory
400 p. 2005 [978-3-540-23953-6]

**Vol. 320:** Steffen, T.
Control Reconfiguration of Dynamical Systems
290 p. 2005 [978-3-540-25730-1]

**Vol. 319:** Hofbaur, M.W.
Hybrid Estimation of Complex Systems
148 p. 2005 [978-3-540-25727-1]

**Vol. 318:** Gershon, E.; Shaked, U.; Yaesh, I.
$H_\infty$ Control and Estimation of State-multiplicative Linear Systems
256 p. 2005 [978-1-85233-997-5]

**Vol. 317:** Ma, C.; Wonham, M.
Nonblocking Supervisory Control of State Tree Structures
208 p. 2005 [978-3-540-25069-2]

**Vol. 316:** Patel, R.V.; Shadpey, F.
Control of Redundant Robot Manipulators
224 p. 2005 [978-3-540-25071-5]

**Vol. 315:** Herbordt, W.
Sound Capture for Human/Machine Interfaces: Practical Aspects of Microphone Array Signal Processing
286 p. 2005 [978-3-540-23954-3]

**Vol. 314:** Gil', M.I.
Explicit Stability Conditions for Continuous Systems
193 p. 2005 [978-3-540-23984-0]

**Vol. 313:** Li, Z.; Soh, Y.; Wen, C.
Switched and Impulsive Systems
277 p. 2005 [978-3-540-23952-9]

**Vol. 312:** Henrion, D.; Garulli, A. (Eds.)
Positive Polynomials in Control
313 p. 2005 [978-3-540-23948-2]

**Vol. 311:** Lamnabhi-Lagarrigue, F.; Loría, A.; Panteley, E. (Eds.)
Advanced Topics in Control Systems Theory
294 p. 2005 [978-1-85233-923-4]

**Vol. 310:** Janczak, A.
Identification of Nonlinear Systems Using Neural Networks and Polynomial Models
197 p. 2005 [978-3-540-23185-1]

**Vol. 309:** Kumar, V.; Leonard, N.; Morse, A.S. (Eds.)
Cooperative Control
301 p. 2005 [978-3-540-22861-5]

**Vol. 308:** Tarbouriech, S.; Abdallah, C.T.; Chiasson, J. (Eds.)
Advances in Communication Control Networks
358 p. 2005 [978-3-540-22819-6]

**Vol. 307:** Kwon, S.J.; Chung, W.K.
Perturbation Compensator based Robust Tracking Control and State Estimation of Mechanical Systems
158 p. 2004 [978-3-540-22077-0]

**Vol. 306:** Bien, Z.Z.; Stefanov, D. (Eds.)
Advances in Rehabilitation
472 p. 2004 [978-3-540-21986-6]

**Vol. 305:** Nebylov, A.
Ensuring Control Accuracy
256 p. 2004 [978-3-540-21876-0]

**Vol. 304:** Margaris, N.I.
Theory of the Non-linear Analog Phase Locked Loop
303 p. 2004 [978-3-540-21339-0]

**Vol. 303:** Mahmoud, M.S.
Resilient Control of Uncertain Dynamical Systems
278 p. 2004 [978-3-540-21351-2]

**Vol. 302:** Filatov, N.M.; Unbehauen, H.
Adaptive Dual Control: Theory and Applications
237 p. 2004 [978-3-540-21373-4]

**Vol. 301:** de Queiroz, M.; Malisoff, M.; Wolenski, P. (Eds.)
Optimal Control, Stabilization and Nonsmooth Analysis
373 p. 2004 [978-3-540-21330-7]

**Vol. 300:** Nakamura, M.; Goto, S.; Kyura, N.; Zhang, T.
Mechatronic Servo System Control
Problems in Industries and their Theoretical Solutions
212 p. 2004 [978-3-540-21096-2]